Shifting Food Facts

This book offers a reframing of food discourse by presenting alternative ways of thinking about the changing politics of food, eating, and nutrition. It examines critical epistemological questions of *how* food knowledge comes to be shaped and *why* we see pendulum swings when it comes to the question of what to eat.

As food facts peak and peril in the face of conflicting dietary advice and nutritional evidence, this book situates shifting food truths through a critical analysis of how healthy eating is framed and contested, particularly amid fluctuating truth claims of a "post-truth" culture. It explores what a post-truth epistemological framework can offer critical food and health studies, considers the type of questions this may enable, and looks at what can be gained by relinquishing rigid empirical pursuits of singular dietary truths. In focusing too intently on the separation between food fact and food fiction, the book argues that politically dangerous and epistemically narrow ideas of one way to eat "healthy" or "right" are perpetuated. Drawing on a range of archival materials related to food and health and interviews with registered dietitians, this book offers various examples of shifting food truths, from macro-historical genealogies to contemporary case studies of dairy, wheat, and meat.

Providing a rich and innovative analysis, this book offers news ways to think about, and act upon, our increasingly complex food landscapes. It does so by loosening our empirical Western reliance on singular food facts in favour of an articulation of contextual food truths that situate the problems of health as problems of living, not as individualistic problems of eating. It will be of interest to students, scholars, and practitioners working in food studies, food politics, sociology, environmental geography, health, nutrition, and cultural studies.

Alissa Overend is an Associate Professor in the Department of Sociology at MacEwan University. Her research in the areas of critical food, health, and nutrition studies questions the shaping of contemporary health and illness phenomena and the politics of knowledge therein.

Critical Food Studies

Series editors: Michael K. Goodman, *University of Reading, UK* and Colin Sage, *Independent Scholar*.

The study of food has seldom been more pressing or prescient. From the intensifying globalisation of food, a world-wide food crisis and the continuing inequalities of its production and consumption, to food's exploding media presence, and its growing re-connections to places and people through 'alternative food movements', this series promotes critical explorations of contemporary food cultures and politics. Building on previous but disparate scholarship, its overall aims are to develop innovative and theoretical lenses and empirical material in order to contribute to – but also begin to more fully delineate – the confines and confluences of an agenda of critical food research and writing.

Of particular concern are original theoretical and empirical treatments of the materialisations of food politics, meanings and representations, the shifting political economies and ecologies of food production and consumption and the growing transgressions between alternative and corporatist food networks.

Hunger and Postcolonial Writing
Muzna Rahman

Digital Food Cultures
Edited by Deborah Lupton and Zeena Feldman

Food Insecurity
A Matter of Justice, Sovereignty, and Survival
Edited by Tamar Mayer and Molly D. Anderson

Shifting Food Facts
Dietary Discourse in a Post-Truth Culture
Alissa Overend

For more information about this series, please visit: www.routledge.com/Critical-Food-Studies/book-series/CFS

Shifting Food Facts

Dietary Discourse in a Post-Truth Culture

Alissa Overend

LONDON AND NEW YORK

First published 2021
by Routledge
2 Park Square, Milton Park, Abingdon, Oxon OX14 4RN

and by Routledge
52 Vanderbilt Avenue, New York, NY 10017

Routledge is an imprint of the Taylor & Francis Group, an informa business

© 2021 Alissa Overend

British Library Cataloguing-in-Publication Data
A catalogue record for this book is available from the British Library

Library of Congress Cataloging-in-Publication Data
A catalog record has been requested for this book

ISBN: 978-1-138-54955-5 (hbk)
ISBN: 978-1-351-00011-6 (ebk)

Typeset in Times New Roman
by codeMantra

I dedicate this book to all those who have ever questioned and resisted singular food truths.

Contents

Figures

Preface

In its earliest germination, this book began a dozen or so years ago. At the time, I was in the thrall of dissertation writing, chipping away at a project that seemed interminable and trying to develop a few coping skills—one of which was cooking. I was drawn to cooking because as I remember saying then: "I need a project that starts and ends in the same day". I saw cooking as a hands-on project that balanced my intellectual work. It was only in the early stages of this book that I realized that the sociology I was doing on my computer and the cooking I was doing in my kitchen were not separate. All the years I thought I was stepping away from the questions that motivate my intellectual pursuits—questions around the politics of knowledge that shape how we come to know a subject or object—I was actually further immersing myself in them. As feminist philosopher Lisa Heldke (2006) upholds, "The various activities of foodmaking can and should stand as models for inquiry precisely because they can puncture the border between theorizing and engaging in practical activity" (p. 216). What I hope this book will offer is just that: a new way of thinking about—and engaging with—food that contemplates the various complex social contexts in which we live and eat.

Acknowledgements

To all the people who helped bring this book to print, my sincerest gratitude for your professional and personal support as well as for your keen eyes, welcoming ears, and critical minds. Thanks also for all the significant and mundane discussions about food, politics, and the increasingly strange times we find ourselves in.

Foremost thanks go to the 17 dieticians who, across geographical and disciplinary divides, were thoughtful, critical, and generous with their time and knowledge and who care deeply about the ongoing effects of food injustice and insecurity. I also want to recognize the friends, colleagues, agencies, and sometimes strangers that connected me with these participants. Even amidst all the formalities of research, it continues to happen in the most unexpected and wonderful ways. I would be remiss if I did not also thank Louise Sicard at Health Canada, who, in the early months of a global pandemic, helped me secure copyright for many of the images featured in this book.

As many scholars and writers will attest, sharing works-in-progress is a deep exercise in trust and vulnerability. I am extremely grateful to have been able to place this trust in the talented, careful hands of Natalie Loveless and Cressida Heyes who read early, multiple, and *much* less-polished versions of the manuscript that follows. With certainty, this project is stronger because of their sustained input—Natalie, for your gentle, never cruel, optimism, and Cressida for your keen, never unwanted, skepticism. I am also indebted to my critical dietetics and food studies colleagues, in particular Maya Hey, Jennifer Brady, Jacqui Gingras, Barbara Parker, Andrea Noriega, Myriam Durocher, and Adele Hite, for creating and maintaining academic communities where conversations outside the confines of the dominant can thrive.

To a group of people I consider mentors for the project, including Alexis Shotwell, Melanie E. DuPuis, Lisa Heldke, Kyla Wazana Tompkins, Charlotte Bitlekoff, Andrea Wiley, Allison and Jessica Hayes-Conroy, John Coveney, and Gyrogy Scrinis: your collective works have been steadfast touchstones in the long, often solitary processes of writing and thinking. Heather Zwicker and Amy Kaler provided much-needed early guidance on the project. My appreciation also goes out to Carolina Cambre who helped spark a very early interest in food studies. Thank you also to Mary Louise Adams and the late Sharon Rosenberg whose careful and thoughtful approaches to scholarship and writing still guide me.

To my friends and colleagues at MacEwan University, including Fiona Angus, Tami Bereska, Shelley Boulianne, Katie Biitner, Michael Gulayets, Randy Jenne, Emily Milne, Margaret Milner, Joanne Minaker, Craig Monk, Andrew Patterson, Peter Puplampu, Sandra Rollings-Magnusson, Melike Schalomon, Michael Seredycz, Sarah Shulist, Stephen Speake, Maria Stecyk, Jeffrey Stepnisky, Diane Symbaluk, and Rob Wiznura, your interest in and support for me and the project has not gone unnoticed. A few colleagues in particular were exceptional, including Susan Raine, Amanda Nelund, Kalyani Thurairajah, Bronwyn Snefjella, Ondine Park, and Melissa Hills. And finally, Sarah Copland who provided immense support in the early days of book writing, which has not been forgotten. To the sociology students at MacEwan University, specifically those enrolled in 290 and 402, thank you for your keenness on nutrition and health—two topics at the centre of this book and dear to my heart—and for constantly (implicitly and explicitly) pushing me towards clarity. Thanks to Delphine Brown, Brenan Molzhan, Claudia Sabourin, and Ashlyn Sawyer, who provided student research support at various stages of this project. I would also like to acknowledge that this research would not have been possible without the structural privileges of a tenure-track job, a one-year sabbatical leave from teaching, and funding support from MacEwan University's Faculty of Arts and Science and the Research Office.

The Routledge team, including Ruth Anderson, Hannah Cassidy, Michael Goodman, Peter Hall, Faye Leerink, Nonita Saha, and Karthik Subramaniam, as well as the blind reviewers at both the proposal and manuscript stage, were all tremendous. Thank you for your professionalism and enthusiasm in the often-invisible labour of academic publishing. Also in my editorial corner was the inimitable Joanne Muzak. Much gratitude for your scrupulous work dotting my i's and crossing my t's and for being the first to read the manuscript in its entirety.

Often the names of the people who support us most are not found in the reference page. Sincerest thank yous go to Christine Peschl, Amy Swiffen, Paul Thorne, Laura Patriquin and family, Carol and Aldyn Overend, Melaina Weiss, Melisa Brittain, Beau Coleman, Donia Mounsef, Nat Hurley, Barbara Sander, Kim Van Bakesh, Lisa Bavington, Breagh Ouellet, Lone Odgers, and QC Gu. I'd also like to thank Karen MacArthur for the much-needed respite away from the city. Other people that deserve mention for reasons mostly outside of the formal work of this project but whose influence on the project can also not be under-valued are Sahana Parameswara, for always meeting me in the realm of what *is* possible; my brother Marcus and his family Wendy, Oliver, and Kaitlyn, for your continued support, especially amidst our shared loss; Elissa Perl, for your kind and expansive heart and for loving me as I am; Mika, my steadfast, four-legged, adventure companion; and Evelyn Hamdon, who, when needed and without fail, reminded me of the political necessity of this work and to always speak my personal truths—both of which helped me immensely in the completion of this project. Lastly, my enduring gratitude goes to my mother, Francesca Overend, for the years of memorable meals and for being my truest teacher in all matters of food.

Abbreviations

CAFO   Concentrated Animal Feedlot Operations
DFC   Dairy Farmers of Canada
DOS   Doctrine of signatures
FNIM   First Nations, Inuit, and Métis
ILO   Intensive Livestock Operations
WHO   World Health Organization

Introduction

Food facts in a post-truth culture

For the many of us dizzily orbiting confusing and contradictory debates about healthy eating, it is unlikely that the controversies that surround the seemingly innocuous question of "what to eat" will be resolved anytime soon. In contemporary Western contexts, deliberations about food abound, from the health benefits of going gluten-free to being vegan, paleo, or locavore. The messages of healthful eating are prodigious in breadth and in the range of mediums through which they circulate: from the news to television talk shows, magazines, government documents, websites, blogs, documentaries, commercials, celebrity opinion, and food packaging, not to mention the unsolicited advice from friends, family members, coworkers, and even, sometimes, near strangers. Combine this with the purposeful confusion of industry buzzwords such as "sustainable", "humanely raised", "natural", and "non-GMO" and the growing scientization of Western understandings of food, and it is easy to see why, in the far-reaching—though sometimes nearsighted[1]—words of Michael Pollan (2008), "We are more confused about how to eat than any people in history" (p. 78). In an age of unprecedented food choice—what Paul Rozin (1976) originally dubbed "the omnivore's dilemma"—growing neoliberal food governance, unknown agriculture and ecological change, and revolutions in information exchange, it is no surprise that many of us are overwhelmed, confused, and (if you are like me) ultimately fed up with the question of, and search for, singular truths about what to eat. If we only look at "the what" or substance of food, we negate the ways in which food and eating are also expansive, complex extensions of our social, cultural, and political worlds. My aim is that this book will provide new ways to think about the commonly polarized framings in and through which the question of what to eat is often posed.

Saturated by an empirical focus on the substance of food, what Allison Hayes-Conroy and Jessica Hayes-Conroy (2013) term "hegemonic nutrition"—models of food that are standardized, reductionist, decontextualized, and hierarchical—I articulate a shift in how we come to know and understand what we eat by shifting dominant, hegemonic orientations and attachments to singular food truths. Following some of the recent calls in critical dietetic, food, and nutrition studies for a much-needed reframing of food discourse, this book, in its broadest sense, moves away from "the what" (or substance) of healthy eating towards critical epistemological questions of *how* food knowledge comes to be shaped and *why* we've seen such pendulum swings when it comes to the question of what to eat. As a sociologist driven by questions of knowledge construction and maintenance, to my mind exploring the "how" and "why" questions tells a more productive and nuanced story about the shifting food landscapes in which we find ourselves. In reorienting the focus from the matter and substance of food to a critical epistemology of healthy eating, this book opens up the terms upon which some food knowledge is rendered more legitimate than others and the effects this has on current conceptualizations of healthy eating. As food facts peak and peril in the face of conflicting dietary advice, contradictory nutritional evidence, and a veritable barrage of opinions (expert and otherwise), I situate shifting food truths through a critical questioning of how the concept of healthy eating comes to be framed and contested, notably amid fluctuating claims of a "post-truth" culture. I evoke the term "post-truth" cautiously yet purposefully. I am not anti-truth or anti-science, nor do I do not support a Trump-style assault on fact. When it comes to singular food truths, however, I am interested in what a post-truth epistemological framework can open up for critical food studies—what kinds of questions might it enable? And what might we gain if we can let go of rigid empirical pursuits of singular dietary truths?

In questioning the dominant, normative discourses of health, Jonathan Metzl (2010) rhetorically asks, "How can anyone take a stand *against* health? What could be wrong with health? Shouldn't we be *for* health?" (p. 1, emphasis in original). In the critical collection, *Against Health*, Metzl (2010) positions discourses of health as prescribing and upholding moralistic positions removed from growing social and structural inequalities and rightly asserts that the current (largely individualized) concept of "health itself is part of the problem" (p. 1). As a means of both disrupting and thinking differently about singular food truths commonly at the core of contemporary Western food debates, following Metzl (2010), I offer an analogous position against singular dietetic truths. How can anyone take a stand *against* truth? What could be wrong with truth? Shouldn't we be *for* truth? In an age of increasingly polarized debates about what to eat (and not eat), the search for and maintenance of singular food truths, for me, is itself part of the problem. Too often, the obsession with singular dietetic truths fails to account for the situational, relational, systemic, and structural contexts that more directly affect health than any one micro- or macronutrient. To be clear then, this book does not offer any clear-cut answers on what to eat. Instead, it provides contextual food truths—ones that examine how shifting food facts come to be shaped, through what discursive terms they are enacted, for whose benefit, and the possibilities that emerge in letting go of them.

Why "post-truth"? Why now? Why food studies?

The now infamous term "post-truth" was initially invoked in the 1990s to describe the one-sided US media coverage of the Gulf War, but it wasn't until 2016, alongside Brexit and the US presidential campaign, that the term reached global notoriety (Poole, 2017). "Post-truth" circulated so extensively that year that Oxford Dictionaries named it the 2016 Word of the Year (Lockie, 2017). In the company of its cousin terms "alternative facts" and "truthiness", post-truth refers to circumstances in which objective facts are less influential in shaping public opinion than appeals to emotion and personal belief (Hodges, 2017). What matters most in the post-truth era "is not whether the claims are true but whether those listening would like those claims to be true" (Lockie, 2017, p. 1). While it is the case that those in positions of power have long evoked what Plato called "the noble lie" (Scruton, 2017), what seems to be shifting in the current post-truth context is the degree and extent to which manipulating truth is considered common practice. As Stewart Lockie (2017) details,

> [p]oliticians may have long been among some of the least trusted members of our societies, but the idea of post-truth politics suggests there is an important qualitative difference between the post-truth politician and the spin-doctor of yore. The post-truth politician does not simply pick-and-choose among relevant facts, offer questionable interpretations or avoid inconvenient questions. The post-truth politician manufactures [their] own facts.
>
> (p. 1)

With increasing links to industry and a deeper muzzling of science, a post-truth world marks a shift from what those in political power "want us to believe as true and what they know is false" (Scruton, 2017). The current information and media wars that circulate on food and healthy eating are also far from outside this charge.

Akin to the strategies used by the tobacco companies in the 1950s and debates about global warming in the 1990s (D'Ancona, 2017; Sismondo, 2017), food politics likewise suffers from collusion between industry and researchers; the purposeful obfuscation of consumer knowledge; product adulteration; and in even more disturbing cases, the outright denial of public harm.[2] While these strategies haven't yet been labelled post-truth from within critical food studies—that's in part the aim of this introduction—a brief review of existing food literature shows that many aspects of food politics meet the criteria for post-truth claims and have for a good two decades.

An early pioneer in the field of food studies, Marion Nestle (2002/2013) offers an enduring study of the industry tactics at work in the nutritional health system, from packaging to education to marketing to lobbying to litigation. She shows that as early as the 1990s there was industry affiliation in the organization and creation of government-issued dietary advice. Not only do food industries liberally fund elected government officials, through hard and soft money as well as gifts in-kind, they also advise directly at the level of public education—from dietary guidelines to health claims on products (Nestle, 2013). In addition to longstanding government lobbying, Nestle further documents how food industries target children through school sponsorships; school lunch programs; and an onslaught of television, internet, and advergames featuring fictionalized characters and product-related toys. More overt tactics to shape and control public food knowledge include the sponsorship of whole academic departments as well as litigation against outspoken critics.[3] McDonald's, the United States Cattlemen's Association, and Chiquita Banana have sued London Greenpeace, Oprah Winfrey, and the Cincinnati Enquirer, respectively, for speaking out against their products. From farm to table, nutritional truths are actively doctored at the hands of food industry, contributing to consumer confusion.

Pollan's (2008) second major book, *In Defense of Food*, also documents the purposeful obfuscation of food knowledge set up to benefit producers rather than consumers. Using food theorist Gyorgy Scrinis's (2002) ideology and paradigm of "nutritionism"—the trend in Western and industrialized nations to understand food as the sum total of its nutritional properties—Pollan (2008) details how the culturally created obsession with healthy eating has reached a point where we hear a great deal about nutrients and very little about food. At first glance, a focus on the contents of food may not seem like a manipulation of dietary truth, but Pollan (2008) contends that hypervigilance regarding nutrients functions to bury any straightforward issues with the foods themselves. Amidst the now everyday language of calories, omega 3s, nutrients, amino acids, antioxidants, and probiotics, science is being used strategically to sell food and reshape it in nutritionist terms to the point where "a notorious junk food [can] pass through the needle eye of nutritionist logic and come out the other side looking like a health food" (Pollan, 2008, p. 53). Under these antics, Pollan (2008) goes on to speculate that "it shouldn't be long before we see chocolate bars bearing FDA-approved health claims" (p. 39). When consumers have pushed for clearer labels on factory-farmed meat or genetically modified products, the industry response has been to add vague or empty signifiers such as "sustainably produced" or "made with natural ingredients" to food packaging. Without any clear sense of what these terms actually represent, they complicate, rather than clarify, consumer queries.

While I am sympathetic to the efforts of food lobbyists such as Pollan and Nestle, who argue for clearer, more accurate food labels, or of popular authors such as Timothy Caulfield (2015), who aims to debunk fad diets, these approaches uphold and maintain the notion of static, singular food truths outside of shifting cultural and relational contexts. In Caulfield's (2015) book *Is Gwyneth Paltrow Wrong about Everything?* he bemoans celebrity culture as part of the broader assault on science and aims to debunk celebrity-driven health claims with what he calls "scientifically informed ammunition" (p. 9). Putting his sexist commentary aside,[4] I find the idea that what celebrities spout is the problem and what scientists uphold is our salvation too simplistic to articulate the complex issues of food and nutrition that have long been regulated by social contexts. I am not defending corrupt, biased, or uninformed research. I think that, at times, bad science is part of the confusion around healthy eating, but I do not think it is the only culprit. As I argue in this book, if we are only concerned with separating food fact from food fiction, we perpetuate the politically dangerous and epistemically narrow idea that there is only one way to eat "healthy" or "right" or "good", which negates a whole host of other ways we can conceptualize food and eating in contemporary contexts. As cultural critic Steven Poole (2017) sums up, "the underlying difficulty of today's polemics about post-truth is that people are talking as though it is always obvious and uncontroversial what is fact and what isn't". Steve Fuller (2018) corroborates the idea that the history of Western philosophy has been steadfast in its questioning of "truth". Debates about food truths are likewise not new, as the history of nutrition abundantly documents. Rather than getting caught up in seemingly endless, circular debates about singular food truths, perhaps we need to accept that food, like many other facets of human behaviour, does not fit cleanly—or uncontroversially—into empirical ways of knowing. Even among experts, any singular truths about healthy eating are at best close approximations.

The myths of nutritional precision

Coining what he calls "the myth of nutritional precision" (p. 24), Scrinis (2008) unpacks "the greatly exaggerated representation of scientists' understanding of the relationship among nutrients, food, and the body" (p. 42). Guised in the quantified language of nutrients and calories, many "facts" about healthy eating turn out to be, instead, highly circumstantial. The most obvious case of shifting food truths was the dietary fat fiasco in the 1980s, which saw low-fat dietary advice proliferate as fats were falsely labelled as "bad".[5] The language of nutritional reductionism often at the centre of contemporary dietary discourse deliberately abstracts the complex relationships between food, illness, and the body, which promotes a misguided, one-dimensional understanding of each of these elements. Depending on the long-term dietary and genetic pattern of the individual in question, the quantity in which they are consumed, and the myriad lifestyle and chemical processes through which they are metabolized, all nutrients can be beneficial or harmful. In short, there is no static or intrinsic truth to a food's healthfulness, or if there is, there are significant methodological barriers to fully and/or accurately measuring it.

Despite the ubiquitous talk about the certainties of healthy eating, the uncomfortable—perhaps inconvenient—truth is that "the entire field of nutrition science rests on a foundation of ignorance [...] about the most basic question of nutrition: what are people eating?" (Pollan, 2008, p. 76). These uncertainties are due to challenges in research design, conduct, and interpretation, where there are significant limitations in the ability to obtain accurate information about the dietary intake of individuals and populations. As Bragg and Nestle (2017) sum up, "humans, unlike experimental animals, cannot be caged and fed controlled diets, a problem that makes research results difficult to interpret" (p. 76). Not only do people routinely underreport their daily food consumptions, we lack the methods and technologies to decipher exactly what people are eating. Pollan (2008) details, for example, that when a person reports they ate a carrot,

> the tabulator consults a USDA database to determine exactly how much calcium or beta-carotene that carrot contained. But because all carrots are not created equal, their nutrient content[s] vary with everything from the variety planted and type of soil it was planted in, to the agricultural system used.
>
> (p. 77)

Suffering from their own inadequacies, these tables fail to account for the complexity of the plant's quality, including whether it was subjected to industrial breeding or monocropping practices. There is increasing evidence that the vitamin and nutrient content of a range of garden crops is diminishing (Davis, Epp, & Riordan, 2004; Evich, 2017) and that this is rarely (if ever) accounted for in standardized nutrition measurements. In short, a carrot is never simply a carrot, and this holds huge implications for how we understand and implement frameworks of healthy eating.

Another major limitation of nutritional studies—in line with the critiques of reductive approaches described thus far—is that they tend to study foods out of the context of diet. Most of us are less likely to eat foods in isolation than we are to eat them in combination with other foods, which greatly affects how they are broken down and metabolized by our bodies. When studying foods too strictly from a nutricentric perspective, the interactive aspects of food combinations and/or the order in which they are eaten are overlooked. Scrinis (2008) coined the term "nutricentric" to refer to "a way of looking at and encountering food as being composed of nutrients, which overwhelms other ways of [...] experiencing food" (p. 46). Traditional diets, including those of humoural medicine detailed in Chapter 1, have long considered food order, timing, and other relational and contextual food factors associated with eating. The French paradox is commonly cited as an example of the importance of understanding foods in the broader cultural patterns of diet where the healthfulness of food is not measured by its empirical substance but in its contextual use: the how, how much, and milieu of food consumption (Rozin, Kabnick, Pete, Fischler, & Shields, 2003). French cuisine is notorious for containing many of the food items currently labelled "unhealthy" in much of the mainstream Western food advice, including white breads, rich cheeses, cream sauces, red wines, and cured meats. However, a typical French meal lasts two to three hours, people rarely eat alone, portion sizes are small, and there are social taboos against second helpings (Rozin et al., 2003). In this example, the food context matters more than the food content and yet, in nutricentric approaches, the former is rarely considered.

Epidemiological data on humans' eating habits focuses more on calorie consumption and less on eating rhythms, timing, frequency of eating and sleep, and their effects on metabolism (Biltekoff, Mudry, Kimura, Landecker, & Guthman, 2014). As Biltekoff et al. (2014) note, calories are abstracted, quantified, and decontextualized measures of food; "the idea that we have to eat calories in order to gain weight tells us little about how the body stores, mobilizes, and utilizes that energy, nor does it tell us how food affects the regulation of these processes" (p. 18). Newer nutritional developments and approaches include situational factors that affect health, such as genetic predisposition, epigenetics, hormone levels, life stage, medications, environmental toxins, and gut bacteria, but these factors are far from the norm in mainstream food research. These insights are not only good examples of the many difficulties of nutritional precision, important for my purposes, they also highlight the danger of over-focusing on singular food truths.

The promise of post-truth for critical food studies

Sparking ire from some scientific communities as yet another assault on science and from the general public as a kind of intellectual relativism gone wrong, I claim the concept of post-truth for critical food studies not for either of these purposes but as a framework for thinking beyond the attachment to the empirical, substance-based food claims that are so commonly at the centre of healthy eating discourses. As Benjamin Tallis (2016) argues, post-positivist academics have much to contribute to the seeming quagmires of post-truth culture because they move the conversation beyond "the kind of 'fact-shaming' that is the negative corollary of the equally futile presentation of facts in the hope of 'reason prevails'" (p. 10). I am skeptical of the epistemological contexts in which singular dietetic truths come to be normalized and what we negate by hyper-focusing on them. In a discipline that still upholds positivist notions of truth and in popular discussions that centralize quantitative understandings of eating, a reframing of singular food truths can shift contemporary food discourses from narrowly defined, empirical-based, and individually-applied concepts to multifaceted and intersectional ones that are understood within broader frameworks of power and knowledge. Responding to Charlotte Biltekoff's (2012) call for critical nutrition studies to provide a new kind of dietary literacy, my framing of plural, contextual food truths aims to

> contribute to a new set of competencies that would enable people to better analyze, evaluate, and create messages related to dietary health. Like media literacy, critical dietary literacy would expand our sense of

literacy by treating dietary reform, dietary ideals, and conversations about dietary health as texts that require analysis.

(p. 186)

A post-truth account of food and nutrition also aligns with Hayes-Conroy and Hayes-Conroy's (2013) pursuit of "diverse nutritions" that "open up imaginative space for nutritional alternatives" (p. 4). In thinking differently about singular, reductive food truths, we might find instead new ways to conceptualize longstanding debates concerning food and health—ones that move us away from individual-based solutions about what we eat and buy to those that better situate food within the complex matrixes of social inequality and oppression, as well as within the politics of knowledge therein.

A move away from the seeming neutrality and objectivity of dietetic food truths can open much-needed relational explorations of food's role in neoliberal capitalism, modern-day colonialism, and the growing chasm between who can and cannot eat well in and amidst the West's misguided doctrine of individualized responsibilities for health. In the context of increasingly polarized debates concerning healthy food facts, questioning the terms through which food facts come to be made raises new questions about incomplete empirical-based food framings that have become too commonplace. Inspired by a Foucauldian understanding of knowledge production, the question, in my mind, is not "what is true?" but "what are the conditions of possibility that make it possible to state what is considered true?" Reorienting the search for and maintenance of singular food truths, in step with Biltekoff's (2012) assertion, "What consumers really need is not ever more information about the nutritional content of food or where it comes from, but a more informed, educated, and critical relationship to dietary health and dietary advice" (p. 172), I ask instead, "What are the terms through which healthy eating is constituted? Which aspects of healthy eating are centralized and which are systematically ignored?" My aim in offering a post-truth approach to critical food and dietetic studies is to diffuse and reorient the search for and reliance on singular food truths towards a contextual, relational food framework (to borrow Lisa Heldke's [2012] useful paradigm) that understands the production of shifting food facts. In the context of the post-truth moment, journalist Vinay Menon reminds us that "by all means, let's go to war on fake news. But let's also understand where the battle must be waged. It's minds, not websites, in need of rattling" (quoted in Hodges, 2017, p. 236). If we can understand how dietetic truths come to be shaped, then we can also understand how they come to be discursively dismantled—a task which, within the ongoing war on knowledge, mounting ecological crisis, and widening social inequalities, seems increasingly difficult *and* increasingly vital.

Overview and organization

This book is theoretically informed and situationally applied. I hope it will appeal to a wide range of readers interested in critical health and food studies. Drawing on Heldke's (2006) assertion that "the unexamined meal is not worth eating" (p. 201), I apply the theory-based questions introduced here and developed more fully in Chapter 1 to the changing advice of food guides in Canada (Chapter 2) and to three food-item case studies: dairy (Chapter 3), wheat (Chapter 4), and meat (Chapter 5). These three items were selected because of their significance to Western food industries and because they have consistently been at the forefront of Western dietary debate and controversy. Of the polarized dietary debates that circulate, these three elicit much scientific debate and public controversy, which makes them especially useful for my purposes here.

My aim in analyzing these particular cases is not to modify anyone's use or consumption of these foods; nor is it to add to the growing cacophony of voices that tell people what to eat. I think we can agree that there is already too much of this kind of dietetic dogma in circulation. I want to move away from individualized doctrines on how to eat towards macro-level analyses of the social, structural, political, cultural, and economic conditions that create healthy eating discourses. In an age of increasingly technical, complex, and contradictory food claims, this project analyzes the epistemological frameworks that uphold singular claims to healthful eating. I reorient the question of healthy eating from one of what to eat to one that questions how we have been taught to think about what to eat, ultimately in an attempt to consider differently and multi-factorially our collective relationships to food (and health) in late modern, neoliberal societies.

Methodology

This project draws from both primary and secondary data to document the range of food information that circulates in contemporary contexts. In Chapter 1, I give an overview of shifting food knowledge from ancient times to modern nutritionism, to trace what Foucault (1977) calls a history of the present. In my subsequent case study chapters, I include both formal and informal dietary sources to map the onslaught of food- and diet-related information that we increasingly face. I draw data from a more recent dietetic archive (post–Second World War to the current moment) that includes formal, publicly available government advice; dietitian and nutritionist teachings, including academic publications; industry claims; evidence from food science (including social science research); and informal and popular opinions. I also look at a range of food-, health-, and dietary-related websites and documentaries.

In addition to an extensive archival review and analysis, I also interviewed 17 registered dietitians on the changing discourses and debates that have surrounded each of the book's three case studies (i.e., dairy, wheat, and meat). Rather than relying on existing dietetic teachings and scholarship, I wanted to hear first-hand professional opinions on the core queries of this project from registered dietitians. I recruited participants through the College of Dietitians of Alberta, Dietitians of Canada, word of mouth, online communities, and snowball sampling. I conducted 17 interviews across western Canada, with 12 of them face-to-face and five via the telephone. I asked participants to define healthy eating, whether they endorse dairy, meat, and wheat as healthy food items, and their professional opinions on the food controversies that have surrounded these items in recent years. My aim in pairing primary and secondary data is to map the range of dietary advice that exists and to trace the specific terms through which singular food truths come to be made, maintained, and legitimized in contemporary discourse.

To code and analyze the data, I focused on repetition of key words and themes, paying attention to variations within and among the assortment of texts, and reading for emphasis and detail (i.e., on how certain themes are being presented) (Tonkiss, 2004). These strategies provided a means of organizing the data without being overly prescriptive. In ways that other methodological approaches do not, critical discourse analysis encourages an inquiry into the prevailing structures of knowledge and power that shape and enable the object or subject in question (Mills, 1997; Wood & Kruger, 2000). Singular food truths are thus analyzed not as reflective of already existing realities but as frameworks in and through which knowledge about healthy eating is actively shaped. Critical discourse analysis is particularly effective for interrogating multiple constellations of truth, which, as Brian Hodges (2017) articulates, is critical in the post-truth context:

> When politicians, news anchors, commentators, social media pundits and yes, leaders in health professions education make "truthy" pronouncements, it is useful to have a tool that can interrogate, not simply what *is true*, but also how things come to be accepted as true. Without a tool to explore the *constructions* of truth, we can only hurl back weak pleas that claims are "not evidence-based". [...] Better that we equip ourselves with tools that allow us to understand how truths are constructed and also how they can be discursively dismantled.
>
> (p. 236)

This book aims to be one such tool. In questioning the discursive terms of singular food truths too often at the core of contemporary health discussions, I open up alternative ways to think about the relational, contextual, and shifting politics of food, eating, and nutrition in increasingly complex and compromised health terrains.

Notes

1 Julie Guthman (2007) takes Pollan (2006) to task for reproducing the individualist, moralist, and fatphobic discourses so commonly at the heart of dietary advice, and in doing so, does little to resist the "healthist" discourses he is in part trying to argue against. Pollan's (2006) response to what he frames "the

omnivore's dilemma" (i.e., the mass industrialization of the food system that is purportedly causing the "obesity epidemic") is not to petition government or some other structural response, but rather, to "eat better", "choose better", and "know better", without adequately examining the systemic barriers that render these "solutions" feasible, accessible, or even desirable. She states that "his answer, albeit oblique, is to eat like he does" (p. 78).

2 The Nestlé infant formula scandal of the 1970s still stands as one of the most egregious cases of unethical food marketing. The company knowingly marketed their infant formula as a safe alternative to breast milk in some of the poorest regions worldwide without adequate warnings that the formula cannot be mixed with local water supplies or be used in smaller portions than what the package recommends, placing many newborns at serious risk for dehydration, diarrhea, pneumonia, and death, and casting undue doubt on centuries of traditional knowledge upholding the benefits of breast milk (Boyd, 2012).

3 The Department of Plant and Microbiology at the University of California, Berkeley signed a $50-million partnership with Novratis, the Swiss agricultural and drug company (Nestle, 2013).

4 In the opening pages of his book, Caulfield (2015) recalls attending a Hollywood movie premiere where he describes how "pencil thin Emma Watson waved in our direction but did not stop. But normal-sized Seth Rogan did" (p. 9). Caulfield offers a needless evaluation and normative critique of Watson's body, perpetuating the sexist idea that women's bodies are somehow on display for the pleasure and/or approval of men. Comments like these appear throughout the book. While the book takes aim at celebrity culture in general, it also picks almost exclusively on female celebrities. In addition to Gwyneth Paltrow—the book's namesake—Katy Perry, Hilary Swank, Jennifer Aniston, and Kim Kardashian are also critiqued for being uninformed in the face of "real" science, negating any animal rights, environmental, and/or humanitarian issues raised by their claims.

5 Gary Taubes (2002, 2008) explains how this low-fat dietary advice confused consumers' understanding of the importance of dietary fats and minimized the risks of excessive carbohydrate intake.

References

Biltekoff, C. (2012). Critical nutrition studies. In J. M. Pilcher (Ed.), *The Oxford handbook of food history* (pp. 172–190). Oxford, UK: Oxford University Press.

Biltekoff, C., Mudry, J., Kimura, A. H., Landecker, H., & Guthman, J. (2014). Interrogating moral and quantification discourses in nutritional knowledge. *Gastronomica: The Journal of Critical Food Studies, 14*(3), 17–26. doi: 10.1525/gfc.2014.14.3.17

Boyd, C. (2012). The Nestlé infant formula controversy and a strange web of subsequent business scandals. *Journal of Business Ethics, 106*(3), 283–293. doi: 10.1007/S10551-011-0995-6

Bragg, M., & Nestle, M. (2017). The politics of government dietary advice. In J. Germov & L. Williams (Eds.), *A sociology of food and nutrition: The social appetite* (4th ed., pp. 75–91). South Melbourne, VIC: Oxford University Press.

Caulfield, T. (2015). *Is Gwyneth Paltrow wrong about everything? When celebrity culture and science clash.* Toronto, ON: Penguin Canada.

D'Ancona, M. (2017). *Post-truth: The new war on truth and how to fight back.* London, UK: Ebury Press.

Davis, D. R., Epp, M. D., & Riordan, H. D. (2004). Changes in USDA food composition data for 43 garden crops, 1950 to 1999. *Journal of American College of Nutrition, 23*(6), 669–682. doi: 10.1080/07315724.2004.10719409

Evich, H. B. (2017, September 13). The great nutrient collapse. *Politico.* Retrieved from https://www.politico.com/agenda/story/2017/09/13/food-nutrients-carbon-dioxide-000511

Foucault, M. (1977). *Discipline and punish: The birth of the prison.* New York, NY: Vintage Books.

Fuller, S. (2018). *Post-truth: Knowledge as a power game.* New York, NY: Anthem Press.

Guthman, J. (2007). Can't stomach it: How Michael Pollan et al. made me want to eat Cheetos. *Gastronomica: The Journal of Food and Culture, 7*(2), 75–79. doi: 10.1525/gfc.2007.7.3.75

Hayes-Conroy, A., & Hayes-Conroy, J. (2013). *Doing nutrition differently: Critical approaches to diet and dietary intervention.* New York, NY: Routledge.

Heldke, L. (2006). The unexamined meal is not worth eating: Or, why and how philosophers (might/could/do) study food. *Food, Culture & Society, 9*(2), 201–219. doi: 10.2752/155280106778606035

Heldke, L. (2012). An alternative ontology of food: Beyond metaphysics. *Radical Philosophy Review, 15*(1), 67–88. doi: 10.5840/radphilrev20121518

Hodges, B. D. (2017). Rattling minds: The power of discourse analysis in a post-truth world. *Medical Education, 51*, 235–245. doi: 10.1111/medu.13255

Lockie, S. (2017). Post-truth politics and the social sciences. *Environmental Sociology, 3*(1), 1–5. doi: 10.1080/23251042.2016.1273444

Metzl, J. (2010). Introduction: Why "against health"? In J. Metzl & A. Kirkland (Eds.), *Against health: How health became the new morality* (pp. 1–11). New York, NY: New York University Press.

Mills, S. (1997). *Discourse.* New York, NY: Routledge.

Nestle, M. (2013). *Food politics: How the food industry influences nutrition and health.* (Revised and expanded tenth anniversary ed.). Berkeley, CA: University of California Press. (Original work published in 2002).

Pollan, M. (2006). *The omnivore's dilemma: A natural history of four meals.* New York, NY: Penguin Books.

Pollan, M. (2008). *In defense of food: An eater's manifesto.* New York, NY: Penguin Books.

Poole, S. (2017, May 18). What's the opposite of post-truth? It's not as simple as "the facts". *New Statesman America.* Retrieved from http://www.newstatesman.com/culture/books/2017/05/what-s-opposite-post-truth-it-s-not-simple-facts

Rozin, P. (1976). The selection of foods by rats, humans, and other animals. *Advances in the Study of Behaviour, 6*, 21–76.

Rozin, P., Kabnick, K., Pete, E., Fischler, C., & Shields, C. (2003). The ecology of eating: Smaller portion sizes in France than in the United States help explain the French paradox. *Psychological Science, 14*(5), 450–454. doi: 10.1111/1467-9280.02452

Scrinis, G. (2002). Sorry, marge. *Meanjin,* 61(4), 108–116.

Scrinis, G. (2008). Functional foods or functionally marketed foods? A critique of, and alternatives to, the category of "functional foods". *Public Health Nutrition, 11*(5), 541–545. doi: 10.1017/S1368980008001869

Scruton, R. (2017, June 10). Post-truth? It's pure nonsense. *The Spectator.* Retrieved from https://www.spectator.co.uk/2017/06/post-truth-its-pure-nonsense/

Sismondo, S. (2017). Post-truth? *Social Studies of Science, 47*(1), 3–6. doi: 10.1177/0306312717692076

Tallis, B. (2016). Living in post-truth: Power/knowledge/responsibility. *New Perspectives, 24*(1), 7–18.

Taubes, G. (2002, July 7). What if it's all been a big fat lie? *The New York Times Magazine.* Retrieved from https://www.nytimes.com/2002/07/07/magazine/what-if-it-s-all-been-a-big-fat-lie.html?pagewanted=all

Taubes, G. (2008). *Good calories, bad calories: Facts, carbs, and the controversial science of diet and health.* New York, NY: Anchor Books.

Tonkiss, F. (2004). Analyzing discourse. In C. Seale (Ed.), *Researching society and culture* (pp. 245–260). Thousand Oaks, CA: Sage Publications.

Wood, L. A., & Kruger, R. O. (2000). *Doing discourse analysis: Methods for studying action in talk and text.* Thousand Oaks, CA: Sage Publications.

1 Western genealogies of healthy eating

From humoural medicine to modern nutritionism

Current scientific understandings of food—measured, marketed, and consumed through the language and framework of micro- and macro-nutrients, calories, and vitamins—are so commonplace that many of us fail to consider that scientized approaches to nutrition are a mere two centuries old. In this chapter I trace the broad epistemological shifts in Western nutritional wisdom from ancient times to modern nutritionism. My aims in showcasing the breadth and variability of food knowledge that has existed within recorded Western histories are twofold. One, by positioning the current nutricentric model as one epistemology among others, I denaturalize the dominant scientific paradigm. Two, in moving away from scientized approaches as *the* singular truth in food and dietetic knowledge, I enable other (post-truth and post-positivist) food paradigms to exist and emerge. This chapter is very specifically *not* an exhaustive history of food or nutrition, if that were even possible or desirable. As Kamminga and Cunningham (1995) note,

> The history of nutrition is an enormous subject that spreads across the history of food, its production and distribution, the history of diet and eating habits, the history of laboratory investigations of animal physiology and of foodstuffs, and the history of society not to mention the history of climate and soil.
>
> (p. 1)

Instead, I offer a genealogy of changing Western conceptualizations of food and nutrition—one that documents the fluctuation of these definitions and the emergence of a dominant, not singular, positivist food paradigm.

Following the intellectual queries set out by Thomas Kuhn (1996) and Michel Foucault (1970, 1975, 1977, 1986), I trace the broad paradigm and epistemological shifts in Western ways of seeing, understanding, and classifying "healthy eating" in order to offer a theoretically informed understanding of what Foucault (1977) famously coined "a history of the present" (p. 31). Using his loosely defined genealogical method, Foucault (1977, 1986) used history as a means of critically understanding the present. As David

Garland (2014) explains, Foucault's historical analysis was not intended to judge historical concepts through contemporary values, nor was it meant to reimagine the past in new ways. These are also not my intentions. Foucault's genealogy was decidedly oriented in the present moment, and, more specifically, on the terms in and through which the present moment has come to be defined. As he (1984) himself expressed, "I set out from a problem expressed in the terms current today and I try to work out its genealogy. Genealogy means that I begin my analysis from a question posed in the present" (quoted in Garland, 2014, p. 367). In this regard, Foucault's histories differ from those written by traditional historians in that they are not concerned with the origin of a particular idea or concept but rather with the epistemic logics that shape its development. Just as Foucault's (1975) history in *The Birth of the Clinic* was concerned with "determining the conditions of possibility of medical experience in modern times" (p. 35), I am interested in understanding the current terms of healthy eating discourse in contemporary culture. I open up the empirical frameworks in and through which modern constellations of healthy eating are judged.

One of the most cited works in the humanities and social sciences, Thomas Kuhn's *The Structure of Scientific Revolutions* (1996) contends that scientific facts are paradigm-relative and change as dominant epistemological paradigms evolve. Rather than facts about the natural world being objectively measured, developed, and/or evaluated, "what man [sic] sees depends both upon what he looks at and also upon what his previous visual conception experience has taught him to see" (Kuhn, 1996, p. 113). In tracing paradigm changes within the natural sciences, Kuhn's work has been influential in moving away from positivist-based accounts of scientific knowledge. His model of shifting paradigms is helpful for understanding how existing scientific ideas are eventually replaced by different ways of thinking. He contends that paradigm changes do not grow out of a development-by-accumulation model, but rather as abrupt changes to the former ways of thinking. Understandably, Kuhn's analysis of paradigm shifts was controversial to the natural sciences at the time because it dispelled the myth about linear, cumulative, and progression concepts of positivist knowledge. However, as Steve Fuller (2000) asserts, it remains to this day one of the few major works in the philosophy of science that has been received sympathetically by natural scientists. Kuhn's theory of paradigm relativity has been used to explain the incommensurability of knowledge across different academic fields, articulating why multiple and sometimes competing facts can coexist. Within different paradigms there can be no common measure between disparate ways of thinking, including about what counts as good science (Friedman, 2003). Given the barrage of conflicting and contradictory food truths that exist within popular and scientific discourses, Kuhn's theory of incommensurability is one way to conceptualize the production of plural food facts.

Like Kuhn (1996), Foucault (1970) was also interested in how knowledge frames shift based on social, historical, and intellectual contexts.[1] However, Kuhn and Foucault's specific interests, as well as how they each went about exploring paradigmatic and epistemological divides, diverge. While Kuhn's (1996) concern was with the natural sciences, Foucault's (1970) application was in the social sciences. Furthermore, while Kuhn focuses on "the shared understandings that bind communities of scientists in social processes of acculturation and replication, Foucault's analyses focus on the often unconscious operation of historically specific epistemological structures" (Garland, 2014, pp. 370–371). Using his earlier archaeological method, which sought to excavate historical ways of thinking, in *The Order of Things* Foucault (1970) traces the epistemic conditions that produce discourse and order thought. Utilizing the language of epistemes—a set of unconscious knowledge structures that govern the discursive formation of objects and subjects—he (1970) contends that "in any given culture and at any given moment, there is always only one episteme that defines the conditions of possibility of all knowledge" (p. 168). His subsequent genealogical work both expands and refracts this earlier statement to articulate a theory of knowledge that highlights how dominant—not singular—epistemic frameworks emerge.

Foucault's later genealogical method articulates that dominant ways of thinking are the result of historical contingencies, not progressive trends in knowledge formation (Gutting, 2014). As its name suggests, genealogy is "a search for processes of descent and emergence" (Garland, 2014, p. 372). Foucault's (1977, 1986) genealogical approach has been widely taken up across a range of social science and humanities disciplines as a way to denaturalize dominant forms of knowledge, including but not limited to scientific, colonial, and patriarchal epistemes. As Gary Gutting (2005) writes, "Foucault's histories aim to show the contingency—and hence surpassability—of what history has given us" (p. 10), which has allowed other (often marginalized) forms of knowledge not only to exist but also to call into question the singularity of dominant truths. Foucault (1984) writes that "the search for descent is not the erecting of foundations: on the contrary, it disturbs what was previously thought to be immobile; it fragments what was thought unified; it shows the heterogeneity of what was imagined consistent with itself" (p. 82). Knowledge and truth are thus contingent.

At a time when narrowly defined, empirical questions of eating dominate discussions of health and encourage didactic, all-or-nothing approaches to food, many of us in the growing fields of critical dietetic, nutrition, and food studies are looking for new ways to reassess prevailing dietary approaches. Drawing on Kuhn's (1996) concept of shifting paradigms and Foucault's (1970, 1975, 1977, 1986) archaeological and genealogical contributions to changing epistemologies, I map Western genealogies of contemporary food knowledge, and the ways we have been taught to think about food and healthy eating. In her work on the history of American dietary advice, Charlotte Biltekoff (2012) asserts "that [a] historical investigation of nutrition and dietary health [...] is exactly what is missing from our public discourse around food and health" (p. 173). Such a discourse would enable new public literacy around the cultural politics of dietary health. Responding to Lisa Heldke's (2006) call for a theoretically informed understanding of food, alongside Biltekoff's desire for a historical food literacy, I trace the macro epistemological shifts in Western nutritional wisdom, highlighting the degree to which changing paradigms (and epistemes) alter, at times vastly, throughout historical Western dietetics. Food knowledge is positioned not simply as a given truth, but as provisional and *actively* produced by shifting paradigms and epistemologies.

Humoural medicine: ancient and Renaissance food knowledge

For more than 15 centuries in much of Europe and its colonies, and currently in some non-nutricentric accounts of food, the dominant discourse of food and nutrition stemmed from theories of humoural medicine.[2] Within these models, food and diet were considered part of a broader, dietetic, and holistic regimen of the self. Food rules were commonly set in the context of the broader topic of "hygiene", which indicated methods for maintaining health by means of diet, exercise, and regulation of all external factors that affect the individual (Albala, 2002). As central as diet may be by contemporary standards, proper dietetic care in ancient and Renaissance periods was considered the single best practice a person could undertake, both for corporeal and intellectual well-being (Skiadas & Lascaratos, 2001). In Plato's words, "There ought to be no other secondary task to hinder the work of supplying the body with its proper exercise and nourishment" (quoted in Skiadas & Lascaratos, 2001, p. 533). Similarly, for Avicenna, a Persian physician and writer, digestion was "the root of life" (quoted in Albala, 2002, p. 54). Pictorius, a Renaissance physician, called the stomach "the *padre di famiglia* [the father of the family]—the supplier of the household, without whom the whole would parish" (quoted in Albala, 2002, p. 54). Patriarchal frames notwithstanding and in contrast to today's largely mechanical understanding of digestion, to ancient and Renaissance thinkers, food affected every aspect of the individual, including, most strongly, the links between diet and health (Anderson, 1997).

According to ancient and Renaissance dietetics, food and health were inextricably linked. Diet was both the cause of disease imbalance and the means of treating ailments of the time. Plato urged that human illnesses should be treated through the regulation of diet before the use of medication: "Wherefore one ought to control all such disease, so far as one has the time to spare, by means of dieting rather than irritate a fractious evil by drugging" (quoted in Skiadas & Lascaratos, 2001, p. 536). In more extreme humoural and pathological imbalances, physical therapies such as hydrotherapies, bloodletting, and cauterizations, as well as drug treatments, were attempted, but only after dietary therapies were exhausted. Diet was the preferred method of treatment because food was considered more similar to human substance and could therefore "be assimilated after digestion and converted into the body" (Cardenas, 2013, p. 261). By contrast, medicine was seen as too distinct from human flesh and while it was able to change the "body's own nature, it could not be converted into the body's own substance" (Cardenas, 2013, p. 261). Diet therapies were used extensively for many generations—and continue to be in some cultural contexts—to maintain and rebalance humours.

Although Hippocrates did not put forth a complete theory of humoural medicine, he is often credited for attributing foods with heating, cooling, moistening, and drying properties, though, as E. N. Anderson (2005) notes, he claims to have inherited the classification method from earlier sources. It was Galen, a Greek physician and disciple of Hippocrates, who advanced and popularized the idea that disease states were the result of an imbalance of the bodily humours necessary for its regulation, maintenance, and function. According to humoural theory, blood was classified as hot and moist, yellow bile as hot and dry, phlegm as cold and moist, and black bile as cold and dry (Estes, 2000). An excess in any of these humours would lead to sanguine, choleric, phlegmatic, and melancholic diseases, respectively (Crowther, 2013). Fluctuations of humours could be the result of temperament, disposition, age, activity level, season, or dietary unevenness. Humoural imbalances could also determine a person's physical characteristics (such as a rosy or pale complexion), mental characteristics (such as a melancholic or bilious personalities), as well as their susceptibility to particular illness (cholera, for example, was presumed to be the result of an excess of yellow bile). The principle behind humoural medicine was to rebalance humours by consuming foods with the opposite properties to the symptoms described by the patient (Crowther, 2013). For example, a physician would attempt to correct phlegmatic conditions or symptoms (i.e., those that were considered a result of an excess of cold and moist properties) with foods that were classified as hot and dry. While the theory of holistic balance was relatively straightforward and widely accepted, the classification of hot/cold and wet/dry foods was more complicated.

Detailed in his book *Eating Right in the Renaissance*, Ken Albala (2002) documents how humoural properties were foremost categorized through taste. The tongue was the first indicator—a kind of litmus test—for effects foods would have on the rest of the body. Black pepper, which was considered hot, would burn or warm the tongue and therefore was presumed to have a similar heating effect as it passed through the body; sour foods, which were classified as cool and dry to the tongue, would have similar constricting effects on the rest of the body; and cooling foods, such as cucumbers, which were fresh and moistening to the tongue, were considered to hydrate the body. Once given a humoural designation along the axes of hot/cold and wet/dry, foods were then subcategorized according to their intensity. First- and second-degree foods mildly affected the body, while third- and fourth-degree foods, such as garlic and ginger, strongly tempered humours and are still used as medicine (Albala, 2002). Differences in individual and community classifications (i.e., what is considered hot for one may be neutral to another) were not only accounted for in the range of dietary advice prescribed, but also provided diagnostic clues to physicians. Counter to many of the broad-scale dietary recommendations seen today, Renaissance phrases such as "*chaçun a son gout* (to each their own tastes)" and "every person in their humor" (Albala, 2002, p. 87) showed the highly individualized approach of humoural dietary theory. Notably, differences in taste or classification were not considered weaknesses to the epistemological system; instead, they were seen as clues that helped prescribe appropriate antidotal remedies to each person based on a number of relational and contextual factors, such as geographical location, season, and family history.

In addition to taste, a food's colour was used to determine its humoural properties. Red and yellow foods, such as bell peppers, were considered heating; green foods, like lettuce or spinach, were considered cooling; and foods pallid in colour, such as rice and bread, were considered to have neutral effects on the body (Anderson, 1997). Interestingly today, some people still eat pallid or bland foods when recovering from an upset stomach as these are considered easier on the digestive system. Food knowledge was also not solely derived from sensorial qualities, such as taste or colour. Another major consideration of humoural designation was the physical environment in which plants and animals grew (Albala, 2002). Marsh plants, for example, were considered phlegmatic (i.e., cool and wet), while mountain plants were considered melancholic (i.e., cool and dry). Furthermore, different parts of the plant (or animal) carried different humoural properties: "Roots, which absorbed the cold, 'undigested' elements directly from the soil, are much harder for us to concoct than hot leaves in which the nutrients have been further refined from the sun" (Albala, 2002, p. 81). Unlike today's homogenous and static food classifications, humoural taxonomies were heterogenous and varied.

Cooking methods, food order, and pairings also played important roles in ancient and Renaissance understandings of food's effects on the body. Potentially damaging substances such as raw meats or eggs were corrected (or balanced) by appropriate cooking methods and by combining balancing foods to counteract deficiencies. The latter is one explanation for why meats, which are commonly heating (or by today's language "pro-inflammatory"), are often combined with vegetables, which are typically cooling, and why heavier red meats were often broken down into soups and stews, rendering them easier to digest (Albala, 2002). Wheat—another food staple food for ancient and Renaissance cultures—also had to be corrected by salt and leavening processes. Without these corrective methods, which are still used today, wheat on its own was too coarse to be effectively digested by the body. While food order was debated at great length by ancient and Renaissance physicians, the general consensus was to start with "opening foods", which is one explanation for why European cuisines tend to start with cooling salads. Jams and cheeses, because of their texture, were consumed to "close the meal" by providing a plug between the stomach and the mouth, and still function in that manner in many European cuisines as desserts (Albala, 2002, p. 99). When considering the intricacy of classificatory factors at play, from individual characteristics and preferences to age and seasonal changes, to the plethora of foods' hot/cold, wet/dry properties, the aim of "balance from all directions" was a daunting task by any generational standard.

By the 19th century, through mass migration and colonization, humoural medicine had spread throughout the globe, blending with the traditional knowledge systems of various cultural groups (Estes, 2000). Humoural medicine and its associated theories of food remain one of the longest-standing documented knowledge systems historically and cross-culturally. As Anderson (2005) notes, "By the mid-20th century, the humoral theory of food was the most widespread belief on earth, far outrunning any single religion" (p. 142). Circulating in some capacity since ancient times (and remember that Hippocrates said he had adopted the system from earlier sources), humoural theory long predates Western empiricism. While much of contemporary Western food culture has moved away from humourism, remnants of this 3,000-year-old system still linger. Many people continue to treat the common cold (the name of the ailment itself a vestige of humoural thinking) with a hot soup, refer to a laid back or "chill" (another word for cold) person as someone who is as "cool as a cucumber", and use the word "hot" as a synonym for spicy (Anderson, 2005, p. 199). Moreover, distant cousins of the humoural system are still widely used in traditional Chinese, Ayurvedic, Indigenous, and some holistic dietary practices, where food and diet are used to counteract disease states. The major epistemological shift in food knowledge that followed—and largely departed from humoural wisdom—was the deeply metaphysical folk concept of the doctrine of signatures (DOS). The DOS emerged out of the spiritual paradigm of the late Middle Ages and circulated as an alternative model to humoural theory into the Renaissance period.

"As above, so too below"[3]: folk medicine and the doctrine of signatures

The guiding premise of the DOS was that the divine creator had endowed "signs-in-nature" (i.e., "signatures", sometimes referred to as sympathies) that pointed healers to the curative potential of foods and plants. Unlike humoural medicine, which focused on a food's taste, colour, and location of growth, the DOS was concerned with its shape and morphological form.

The DOS theory contended that a food's visual structure provided clues to the body part or ailment it was intended to heal (Richardson-Boedler, 1999). A commonly cited example is the walnut, which resembles the human brain and was widely promoted to treat head ailments (Panese, 2003). William Coles, a 17th-century botanist describes, "The Kernel hath the very figure of the brain, and therefore it is very profitable for the brain" (quoted in Pearce, 2008, p. 51). Similarly, there existed symmetry between the aconite flower and human eyes. The 16th-century alchemist Oswald Croll explains that "the sign [of aconite] is easily legible in its seeds: they are tiny dark globes set in white skin [*sic*] like coverings whose appearance is much like that islands covering an eye" (quoted in Foucault, 1970, p. 27). While the theory of the DOS had circulated orally in various cultures since Antiquity, it was only in the 16th and 17th centuries that scholars, philosophers, and physicians began to collect, consolidate, and document the body of knowledge that comprised the doctrine (Lev, 2002).

A number of European scholars, including pioneers in modern toxicology and botany, were attracted to the DOS (Bennett, 2007). One of the earliest proponents of the DOS was Paracelsus, a 16th-century Swiss physician and alchemist who critiqued earlier Hippocratic and Galenic teachings. He believed that humoural theory was too limited to account for the scope and complexity of human ailments and insisted instead that health was achieved by maintaining unison with the heavens (Richardson-Boedler, 1999). For him and other supporters of the DOS, the essence of all things became apparent by studying their material form, which was endowed in all of nature (Lev, 2002). In his words, "There is no thing in nature, created or born, which does not strive to reveal its inner form outwardly; for the inner life continually works toward revelation" (quoted in Richardson-Boedler, 1999, p. 174). Reinforcing Paracelsian theory was Jakob Böhme, a 17th-century German mystic and theologian, who alleged that "the greatest understanding lies in the signatures, wherein man may not only learn to know himself, but also the essence of all essences" (quoted in Bennett, 2007, p. 248). Food, akin to the many wonders of the natural world, was a microcosm of the divine. Human bodies and their ailments were likewise considered extensions of the divine and the natural world, which were considered one and the same. Guided by the spiritual beliefs of the epoch, the divine was everywhere in the living world—"one had only to look closely enough" (Lev, 2002, p. 14).

The DOS formalized the belief of similitude between humans and the pre-classical world they inhabited. In it, food and everything else of the natural world was held together by a universal, cosmic chain of symmetry. Foucault (1970) explains how the DOS, as an example of the pre-classical episteme, articulated an ontological framework where "the universe was folded in upon itself: the earth echoing the sky, faces see themselves reflected in the stars, and plants holding within their stems the secrets that were of use to man" (p. 17). Humans were not separate from the material or spiritual worlds they inhabited, nor was the food they consumed. For Paracelsus, humans and the physical world were "two twins who resemble one another completely, without it being possible for anyone to say which of them brought its similitude to the other" (quoted in Foucault, 1970, p. 20). Likewise, for Croll, humans were nested in a succession of symmetry, analogy, and emulation: "His flesh is in glebe, his bones are rocks, his veins great rivers, his bladder is the sea, and even his seven principal organs are the metals hidden in the shafts of mines" (quoted in Foucault, 1970, p. 22). Knowledge and the heavens were fused and epochal understandings of food were merely an extension of this thinking.

From the late Middle Ages to the 19th century, the DOS circulated—and in some cases flourished—as an alternative view to humoural theory. While Galen and Hippocrates subscribed to the healing epistemology of antipathy (i.e., opposites cure), Paracelsus and his followers espoused the healing philosophy of sympathy (i.e., like cures like) (Panese, 2003). The DOS first gained traction in its use of curative plants, but as support for the doctrine grew, so did its application. The doctrine eventually widened to include animal-, mineral-, and chemical-based medicines (Richardson-Boedler, 1999). Paracelsus, again a pioneer in this area, initiated the inclusion of metals in the early creation of pharmacopeia. In his now famous dictum, "The dose makes the poison", he was the first to propose that all substances are potentially poisonous (Richardson-Boedler, 1999, p. 174). His discovery that it is the quantity not the substance that renders an object hazardous to human health was foundational to the current fields of toxicology, immunology, and homeopathy (Pearce, 2008).[4]

As in the case of humoural medicine, there are remnants of the DOS still in circulation. Ginger root, whose shape resembles the human stomach, continues to be widely used as a digestive aid. The walnut, resembling the human brain and high in omega 3s, is known to help brain function. A carrot, when sliced and held up to the light, bears a stunning resemblance to the human eye and is known to promote healthy vision due to its high levels of beta carotene. Despite these lingering examples, as a broad theory of food, diet, and healing, the DOS was eventually debunked. As early as 1583 Rembert Dodoens, the 16th-century Flemish physician and botanist, pointed out that the DOS "is so changeable and uncertain that it almost seems unworthy of acceptance" (quoted in Bennett, 2007, p. 249). In short, there was far too much interpretation about what constituted similarity for the theory to be universally accepted. In predominantly illiterate societies, the DOS was most likely used as a mnemonic method for recalling and classifying a wide range of curative plants (Panese, 2003). And, in a highly spiritual paradigm, the DOS was "rather fancied by men than designed by Nature", understood in today's terms as a kind of confirmation bias (Ray, 1717, quoted in Bennett, 2007, p. 251). Despite the epistemic divides between the DOS and the incumbent view of modern nutritionism, sight continued to play a constitutive role in dietary theory. What changed through the paradigm of modern nutritionism, however, was *how* one came to see food and, correspondingly, *what* came to be seen.

Modern nutritionism

Gyrogy Scrinis (2008, 2013) coined the term "nutritionism" to refer to the overly reductive approach to food that derives from, but is not synonymous with, nutritional science. While the concept of nutritionism overlaps with aspects of nutritional science, the two are not equivalent. Emphasizing the division between them, Scrinis (2013) articulates the following:

> It is [...] important to make clear that nutritionism and nutritional reductionism [...] do not simply refer to the study or understanding of foods in terms of their nutrient parts. If this were the case, then all scientific research into nutrients, and all nutrient-specific dietary advice, would be necessarily reductive. Rather, it is the *ways* in which nutrients have often been studied and interpreted, and then applied to the development of dietary guidelines, nutrition labeling, food engineering, and food marketing, that are being described as reductive.
>
> (p. 5, emphasis in original)

The ideology of nutritionism, then, can be understood as a narrow operationalization of scientific food frameworks. In what follows, I map some of the scientific and empirical epistemologies that led to the spread of modern nutritionism as the dominant Western framework of contemporary food knowledge.

Commonplace by contemporary Western standards, scientific understandings of food date back to the chemical revolution in France at the end of the 18th century. The identification of chemical properties and the development of methods of chemical analysis led to quantitative ideas concerning food and how food was used by the body (Carpenter, 2003). In 1827, summing up the work of chemists of the past three decades, the 17th-century English biochemist William Prout divided foods into three substances: the saccharine, the oily, and the albuminous (i.e., resembling animal protein). These classifications would later come to be reclassified as carbohydrates, fats, and proteins, respectively, and would form the basis of a macronutrient approach to food (Kamminga & Cunningham, 1995). Entering what historian Harvey Levenstein (1988) refers to as the era of new nutrition, scientists were concerned with the nutritional and chemical elements of foods and with the effects these could have on the human body. Scientific understandings of food, as Stephen Mennell (1987) argues, also guised colonial attitudes about how and what to eat, where the language of macronutrients became central tropes in what he terms the "civilizing of appetite" (p. 373), designed to presumably improve traditional diets. Food was no longer understood by its humoural or morphological characteristics, but instead by its internal nutrient properties. Concerned with how these newfound chemical properties were metabolized by the body, 19th-century chemists Justus von Liebig and Wilbur Atwater were among the early pivotal figures in the changing views of human nutrition (Lupton, 1996).

In her book *Measured Meals*, Jessica Mudry (2009) explains that Liebig developed a quantitative approach to nutrition from his studies in inorganic chemistry. In the 1830s, he began to analyze the chemical makeup of living organisms. Liebig saw similarities between plant and animal metabolism and promoted the idea that nitrogenous animal matter was similar to and derived from plant matter and, as such, could be used to build animal tissue. Alongside early industrialization, Liebig, who was also an entrepreneur, became singularly focused on the value of protein for its capacity to invigorate the body and promote muscular growth (Carpenter, 2003). Upholding protein as "the only true nutrient", he created, marketed, and sold "Liebig's Extract of Meat" (i.e., the precursor to OXO meat extracts), which achieved wide commercial success in no small part due to his overemphasis of the value of protein in the human diet (Scrinis, 2013, p. 54). Carl von Voit, a 19th-century German physiologist and dietitian, would eventually temper Liebig's narrow protein focus by advocating for a balance among meat, vegetables, and grains. Nonetheless, Liebig's understanding of the chemistry of food was foundational in the shift away from humoural medicine and towards a quantification of dietary knowledge (Kamminga & Cunningham, 1995).

The next building block in the quantification of diet was the small but immeasurable force of the calorie. Derived from the French word "*calor*", meaning heat, the unit of the calorie was used to measure the energy contained in food and burned by the body (Hargrove, 2006). By the end of the 19th century, European and American nutrition scientists, led by the work of Atwater, began studying the energy content of various foods and the amount of energy expended during a range of activities (Scrinis, 2013). Using a calorimeter, Atwater extensively measured and documented the caloric composition of food and which ones best maximized human output.[5] One part of this focus was on which foods yielded the best energy to cost ratios; as Atwater (1902) itemizes,

> [t]en cents spent for beef sirloin at 20 cents a pound buys 0.5 pounds of meat, which contains 0.08 pound of protein, 0.08 pound of fat, and 515 calories of energy, actually available to the body, while the same amount spent for oysters at 35 cents a quart would buy little over half a pound of oysters, containing 0.03 pound of protein, 0.01 pound of fat, 0.02 pound of carbohydrates, and 125 calories of energy; or if spent for cabbage, at 2¼ cents a pound, it would buy 4 pounds, containing 0.05 pound of protein, 0.01 pound of fat, 0.18 pound of carbohydrates, and 460 calories of energy, while of wheat flour at 3 cents a pound it would buy 3½ pounds, containing 0.32 pound of protein, 0.03 pound of fat, 2.45 pounds of carbohydrates, and 5140 calories of energy.
>
> (Quoted in Mudry, 2009, pp. 40–41)

In a relatively short period of time, a good diet, which was once understood as a matter of balance broadly defined, became "precisely quantifiable through the statistical truths of mathematics" (Turner, 1982, p. 258). The idea of food as fuel, introduced decades prior with the concept of macro-nutrients, was now widely operationalized through the newfound empirical equation of the calorie.

As transformational as the caloric model of food was, however, it failed to account for the persistence of scurvy and other illnesses that continued to plague Europe and North America at the turn of the 20th century, which physicians and scientists long suspected were rooted in diet (Scrinis, 2013). In 1912 the Polish biochemist Casimir Funk hypothesized that beriberi, pellagra, scurvy, and rickets were caused by yet-to-be-identified food de-ficiencies (Carpenter, 2003). He went on to propose that these nutritional deficiencies were caused by a lack of vital amines, which he shortened to "vitamines", later altered to "vitamins" since not all of them were amines (Kamminga & Cunningham, 1995). For the next 30 years, beginning with Elmer McCollum's work on "accessory food factors" A and B (later renamed vitamins A and B), vitamins including riboflavin, folic acid, and vitamin D were the central focus of nutritional research and both replaced and chal-lenged the prior singular focus on the calorie (Cannon, 2002; Carpenter, 2003). Biltekoff (2012) explains that vitamins' "link to deficiency diseases undermined the logic of early nutrition by revealing the fact that calories alone were not enough to promote health and sustain life" (p. 176). Not un-like the divine properties associated with food in the DOS, "these invisible and otherwise inert micronutrients [became invested] with almost magical properties" (Scrinis, 2013, p. 64). Through much of the first half of the 20th century, vitamins were hailed as protective agents against disease as well as for their broader promises of health. "Vitamania", alongside broader em-pirical accounts of nutrition, was shifting the public's understanding and consumption of food (Apple, 1996).

In a matter of a couple hundred years, the dominant food paradigm of Enlightenment Europe had swung from holism to mechanism, from individ-ualization to homogenization, from localization to standardization, from community- to expert-driven, and from one largely concerned with quality to one inherently focused on quantity. What was once fluid, contingent, and complex became increasingly mechanistic—"ordered, controlled, and un-derstood though measurable factors" (Mudry, 2009, p. 2). It was through these epistemic shifts that the ideology of nuritionism emerged and circu-lated as an arm of nutritional science. The rationalization of diet gave way to the corollary rationalization of the human body. As such, human bodies became the perfunctory receptor of food's fuel and a "complex system of pipes, pumps and canals [that] can only be satisfactorily maintained by the correct input of food and liquid, appropriate exercise and careful evacua-tion" (Turner, 1982, p. 260). While diet in ancient and Renaissance societies was once considered part of a holistic practice of the self, and in spiritual settings as an extension of the cosmos, in the new model of scientific nu-trition, food was functional—claiming to fight growing rates of chronic

illnesses that had (for the most part) replaced deficiency diseases of generations prior, while also serving the industrial, colonial, and capitalist projects
of the 20th century.

John Coveney (2000) documents how nutritional science became a central
discourse in the management of modern societies. He explains that

> nutrition strategies engender a form of control which is scientistic—
> where a population is encouraged the adopt scientific conduct in regard
> to food based upon assumptions that it is a 'sick population' and, as
> such, everyone is in need of dietary reform.
>
> (Coveney, 2000, p. 15)

At a time when scientific knowledge about food was growing, so too were
food- and disease-related anxieties, both for the upper and working classes
of Europeans (Mennell, 1985). In her American historical analysis, Biltekoff
(2012) corroborates that scientific framings of food were at their root linked
to the regulation of wider social problems. She details that

> for early nutritionists like Atwater, laboratory findings about the chem-
> ical composition of food were inseparable from the aim of raining in
> impractical excesses among urban workers, reducing poverty, and con-
> tributing to a much-needed improvement in character, morality, and
> social order.
>
> (Biltekoff, 2012, p. 173)

Kathleen LeBesco (2004) and Julie Guthman (2011) also uphold that the
empirical language of modern nutrition disguises social judgements about
what and how much to eat, the victim-blaming imperative of "eating well",
and long-constructed classist values of temperance and self-control, which
are only heightened in healthist, neoliberal contexts. Modern nutritionism
marries scientific approaches to understanding food with social discourses
of healthy eating, not only negating other ways of knowing food, but also
streamlining dominant definitions of healthy food.

Shifting epistemologies of healthy eating

By tracing the broad shifts in the historic framings of food knowledge,
this chapter sets up the ways that dietetic discourse is far from continuous
and has changed—at times radically—between paradigms. The language
of nutrients, calories, and vitamins, while virtually ubiquitous by today's
Westernized food standards, was unknown to past populations. Likewise,
the holistic, descriptive humoural understandings of food have for the most
part been replaced by measured, empirical accounts of eating. Under a nutricentric paradigm, "a cool, creamy glass of milk became 150 calories and
a measurable ratio of fat, proteins, and carbohydrates" (Mudry, 2009, p. 22).

In the epistemic shifts from humourism to nutritionism, the "highly individualistic humoral model was eventually supplanted by dietary guidelines that provided absolutes at the level of the population" (Lupton, 1996, p. 72). These dietetic absolutes have yielded benefits and drawbacks.

On one hand, the search for dietetic universalism through a positivist, empirical frame has brought about significant nutritional insights, including a greater understanding of the role of micro- and macronutrients in the human diet and their roles in disease protection. On the other, an overly nutricentric approach to food and nutrition also bears many shortcomings, including the marginalization and delegitimization of traditional cultural food knowledge and an erasure of food and nutrition's economic, social, cultural, political, familial, and ecological contexts. While nutricentric understandings of food worked well to mitigate the deficiency diseases of the early 20th century, the same model does not equally apply to the many chronic health concerns affecting Western societies in record numbers today (Mayes & Thompson, 2014). The increase, not decrease, in diet-related diseases in the 21st century indicates a failure of a strictly nutricentric paradigm, one that in its biocentric narrowness cannot account for the *social* conditions affecting human health.

In decentering modern nutritionism, I take up dietary discourses not as singular empirical truths, but as products of history shaped by shifting epistemologies. The plurality of food truths seen today can thus be reframed as a strength and not merely an accusation of misguided truth. Plural food truths enable a reconfiguring of our relationships with food as extensions, refractions, and resistance to nutricentric paradigms. In the chapters that follow, I reframe questions of healthy eating beyond singular, nutricentric food truths. In exploring the shifting epistemologies that make up competing truth claims, my goal is to expand the current singular framings of food and nutrition in order to explore how we can complicate the dominant stories that circulate about food, nutrition, and health. A critical contextualization of shifting food truths can address the politics of knowledge and relations of power that make up healthy eating discourses. Amid what Julianne Cheek (2008) calls the "new conservatism" of healthist ideology, where structural health factors are often ignored in the face of individualized responsibility, my critiques aim to counter the individualistic strategies that are part and parcel of nutricentric thinking.

Moving the conversation from one of inherent or intrinsic food truths to one concerned with the shifting contexts of food truths, in the subsequent chapters on food guides, dairy, wheat, and meat I ask, "What conditions of possibility for knowledge about food exist? How do these food frameworks contribute to competing and sometimes contradictory food truths? How have discourses of nutritionism been normalized in the creation and maintenance of modern food knowledge? What dominant epistemic structures have been created, normalized, and perpetuated through the stories we tell about contemporary foods and the debates concerning their healthfulness?" While I am not proposing that we return to ancient humoural food philosophies or to the DOS of the Middle Ages, like Geoffrey Cannon (2002), I suggest that we may want to resurrect one piece of nutrition's long history: the ability to value it as a science *as well as* a philosophy. As my genealogy of dietetic knowledge has shown, how we think about food will alter our relationship with it. After over 200 years of nutritional science as the dominant food paradigm, part of why we are seeing conflicting and sometimes contradictory food truths is because we are starting to see new food paradigms emerge—ones that value context as much as content and ones that look at food relationally instead of strictly empirically. What a genealogy enables is a precisely a "revaluing of values" (Garland, 2014, p. 372) and this is much needed at this particular historical moment.

Notes

1 Given that *The Structure of Scientific Revolutions* was originally published in 1962 and raises parallel questions to Foucault's early interests in the archeology of knowledge, I find it curious that Foucault (1970) does not cite Kuhn's work. As Garland (2014) remarks, Foucault notoriously avoided elaborate citation of authors whose ideas had influenced his work, which perhaps explains the omission. It is also possible that Foucault's (1970) oversight in referencing Kuhn's work was due to the language barriers that existed at the time between the two texts. Notably, Gary Gutting (2003) launches a similar critique at Kuhn, who he says failed to draw on the French philosophy of science tradition that was well known at the time Kuhn wrote *The Structure of Scientific Revolutions*—an oversight, which again, is possibly explained by substantive language barriers.
2 Traditional Chinese, Ayurveda, Indigenous, and some naturopathic medical practices continue to be based on the ancient philosophies of humoural medicine. These practices typically draw on holistic understandings of healing, food as medicine, and the theory of balancing hot/cold and wet/dry elements originating in humoural theory (Anderson, 1997, 2005; Crowther, 2013).
3 "As above, so too below" is an ancient Hermetic saying, shortened from "that which is Below corresponds to that which is Above, and that which is Above, corresponds to that which is Below", which articulated the belief of symmetry between macrocosmic and microcosmic worlds (Wikipedia, 2018).
4 Samuel Hahnemann, the early 19th-century German physician and founder of homeopathy, defined the practice based on the Law of Similar and drew extensively from Paracelsus's work (Richardson-Boedler, 1999).
5 The calorimeter was a machine initially used in the physical sciences to measure the heat of chemical reactions. Atwater effectively used it to render the calorie a measure of human expenditure. Mudry (2009) argues that "for every food that the calorimeter burned, the instrument spoke to the scientist through its heat readings and ash analysis [...] communicat[ing] information about things like motion or protein content and establish[ing] [them] as natural fact[s]" (p. 38).

References

Albala, K. (2002). *Eating right in the renaissance*. Berkeley, CA: University of California Press.
Anderson, E. N. (1997). Traditional medical values of food. In C. Counihan & P. Van Esterik (Eds.), *Food and culture: A reader* (pp. 80–91). New York, NY: Routledge.

Anderson, E. N. (2005). *Everyone eats: Understanding food and culture*. New York, NY: New York University Press.

Apple, R. (1996). *Vitamania: Vitamins in American culture*. New Brunswick, NJ: Rutgers University Press.

Bennett, B. C. (2007). Doctrine of signatures: An explanation of medicinal plant discovery or dissemination of knowledge? *Economic Botany, 61*(3), 246–255.

Biltekoff, C. (2012). Critical nutrition studies. In J. M. Pilcher (Ed.), *The Oxford handbook of food history* (pp. 172–190). Oxford, UK: Oxford University Press.

Cannon, G. (2002). Nutrition: The new world disorder. *Asia Pacific Journal of Clinical Nutrition, 11*, S498–S509.

Cardenas, D. (2013). Let not thy food be confused with thy medicine: The Hippocratic misquotation. *European Society for Clinical Nutrition and Metabolism, 8*(6), 260–262.

Carpenter, K. (2003). A Short History of Nutritional Science: Part 1 (1785–1885). *The Journal of Nutrition, 133*(3), 638–645.

Cheek, J. (2008). Healthism: A new conservatism? *Qualitative Health Research, 18*(7), 974–982. doi:10.1177/1049732308320444

Coveney, J. (2000). *Food, morals and meaning: The pleasure and anxiety of eating*. New York, NY: Routledge.

Crowther, G. (2013). *Eating culture: An anthropological guide to food*. Toronto, ON: University of Toronto Press.

Estes, J. W. (2000). Food as medicine. In K. F. Kiple & K. C. Ornelas (Eds.), *The Cambridge world history of food* (Vol. 2, pp. 1534–1553). New York, NY: Cambridge University Press.

Foucault, M. (1970). *The order of things: An archaeology of the human sciences*. New York, NY: Random House.

Foucault, M. (1975). *The birth of the clinic: An archaeology of medical perception*. New York, NY: Vintage Books.

Foucault, M. (1977). *Discipline and punish: The birth of the prison*. New York, NY: Vintage Books.

Foucault, M. (1984). Nietzsche, genealogy, history. In P. Rabinow (Ed.), *The Foucault reader* (pp. 76–100). New York, NY: Pantheon Books.

Foucault, M. (1986). *The care of the self: Volume III of the history of sexuality*. New York, NY: Pantheon Books.

Friedman, M. (2003). Kuhn and logical empiricism. In T. Nickles (Ed.), *Thomas Kuhn* (pp. 19–44). Cambridge, UK: Cambridge University Press.

Fuller, S. (2000). *Thomas Kuhn: A philosophical history of our times*. Chicago, IL: University of Chicago Press.

Garland, D. (2014). What is a "history of the present"? On Foucault's genealogies and their critical preconditions. *Punishment & Society, 16*(4), 365–384.

Guthman, J. (2011). *Weighing in: Obesity, food justice, and the limits of capitalism*. Berkeley, CA: University of California Press.

Gutting, G. (2003). Thomas Kuhn and the French philosophy of science. In T. Nickles (Ed.), *Thomas Kuhn* (pp. 45–64). Cambridge, UK: Cambridge University Press.

Gutting, G. (2005). *The Cambridge companion to Foucault* (2nd ed.). Cambridge, UK: Cambridge University Press.

Gutting, G. (2014). Michel Foucault. In E. N. Zalta (Ed.), *The Stanford encyclopedia of philosophy* (Winter 2014 ed.). Retrieved from https://plato.stanford.edu/archives/win2014/entries/foucault/

Hargrove, J. L. (2006). History of the calorie in nutrition. *The Journal of Nutrition, 136*(12), 2957–2961.

Heldke, L. (2006). The unexamined meal is not worth eating: Or, why and how philosophers (might/could/do) study food. *Food, Culture & Society, 9*(2), 201–219.

Heldke, L. (2012). An alternative ontology of food: Beyond metaphysics. *Radical Philosophy Review, 15*(1), 67–88.

Kamminga, H., & Cunningham, A. (1995). *The science and culture of nutrition, 1840-1940.* Atlanta, GA: Editions Rodopi, B.V.

Kuhn, T. S. (1996). *The structure of scientific revolutions* (3rd ed.). Chicago, IL: University of Chicago Press.

LeBesco, K. (2004). *Revolting bodies?: The struggle to redefine fat identity.* Amherst, MA: University of Massachusetts Press.

Lev, E. (2002). The doctrine of signatures in the medieval and ottoman levant. *Vesalius, 8*(1), 13–22.

Levenştein, H. A. (1988). *Revolution at the table: The transformation of the American diet.* Berkeley, CA: University of California Press.

Lupton, D. (1996). *Food, the body, and the self.* London, UK: Sage Publications.

Mayes, C., & Thompson, D. B. (2014). Is nutritional advocacy morally indigestible? A critical analysis of the scientific and ethical implications of "healthy" food choice discourse in liberal societies. *Public Health Ethic, 7*(2), 158–169.

Mennell, S. (1985). *All manners of food: Eating and taste in England and France from the middle ages to the present.* Chicago, IL: University of Illinois Press.

Mennell, S. (1987). On the civilizing of appetite. *Theory, Culture & Society, 4*(2–3), 373–403.

Mudry, J. (2009). *Measured meals: Nutrition in America.* Albany, NY: State University of New York Press.

Panese, F. (2003). Doctrine des signatures et technologies graphiques au seuil de la modernité. *Gesnerus, 60*(1–2), 6–24.

Pearce, J. M. (2008). The doctrine of signatures. *European Neurology, 60*(1), 51–52.

Richardson-Boedler, C. (1999). The doctrine of signatures: A historical, philosophical and scientific view. *British Homeopathic Journal, 88*(4), 172–177.

Scrinis, G. (2008). Functional foods or functionally marketed foods? A critique of, and alternatives to, the category of "functional foods." *Public Health Nutrition, 11*(5), 541–545. doi:10.1017/S1368980008001869

Scrinis, G. (2013). *Nutritionism: The science and politics of dietary advice.* New York, NY: Columbia University Press.

Skiadas, P. K., & Lascaratos, J. G. (2001). Dietetics in ancient Greek philosophy: Plato's concept of healthy diet. *European Journal of Clinical Nutrition, 55*(7), 532–537.

Turner, B. S. (1982). The government of the body: Medical regimens and the rationalization of diet. *The British Journal of Sociology, 33*(2), 254–269.

Wikipedia. (2018). Hermeticism. Retrieved March 24, 2018, from https://en.wikipedia.org/wiki/Hermeticism#"As_Above,_So_Below

2   Shifting food facts of Canada's Food Guides, 1942–2019

When I first learned about Food Guides in my elementary school class-rooms, I didn't realize that knowledge about food was so malleable, nor so informed by the companies who made the foods we were encouraged to avoid. While the heyday of the authority of Canada's Food Guides may have been before the internet, social media, and popular food documen-taries, the document remains a key feature in shaping nutritional advice in contemporary culture, even if, as E. Melanie DuPuis (2007) notes, most people do not strictly follow its advice. While not firmly or consciously followed, Food Guides remain a key document in the shaping of healthy eating discourses. Invariably, when teaching this content to my Sociology of Food and Nutrition classes, students will say they do not follow Food Guide advice, and yet when asked, "What constitutes healthy eating?", they will go on to recite near-verbatim Food Guide messaging. Available in a dozen languages, Canada's 2007 Guide to Healthy Eating is the most pop-ular page on the Health Canada website and is the federal government's second most requested print document (Health Canada, 2007a; Hui, 2017). In the three-quarters of a century that food rules and guides have been in circulation, they have been instruments of population control, regulating the health of the nation through individual food choice (Mosby, 2014). They have also undergone name changes, image shifts, and message alterations—at times, significant ones. Initially concerned with mitigating widespread malnutrition, the guide's aim has shifted to countering chronic diseases of adulthood, such as type 2 diabetes, heart disease, and cancer, by target-ing individual food choice (Health Canada, 2007b). Not only have health contexts changed since the initiation of early Food Guides, but, impor-tantly, so too have the social, political, and economic settings in which their advice circulates. Since inauguration of the Official Food Rules in 1942, Canadians have increasingly moved from being producers to consumers of food, alongside the growth of capitalism, the rise of neoliberalism, changes in the structure and organization of the family, and revolutionary changes in information exchange.

While there is a growing body of critical literature on US Food Guides (Biltekoff, 2012; Mudry, 2009; Nestle, 2013), much less academic analysis has been conducted on the Canadian versions, with the exception of Ian Mosby's (2014) historical account of early Food Guides and a few individual pieces that analyze Canadian Food Guides in comparison to international models (Arceño, 2016; Bragg & Nestle, 2017; Murphy & Barr, 2007; Painter & Rah, 2002). Within popular media, and some medical sources, however, there have been mounting critiques about Canada's Food Guides, which I will summarize below. There are similarities between Canadian and international Food Guides, in terms of their food classification, visual representation, and recommended daily intakes, pointing to the dominant discourses that shape public thought about healthy eating. Such parallels also emerged, as Mark Arceño (2016) documents, after the international gathering on food-based dietary guidelines held in Rome in 1992, when many international Food Guides made relatively abrupt changes to dietary guidance. There are also key differences between Canadian and international guides, from specific dietary advocacy to a relatively recent focus on First Nations, Inuit, and Métis nutrition. This chapter provides a brief historical overview of the eight editions of the guide from 1942 to 2007, outlines their critiques, and documents the most recent changes to Canada's healthy eating strategy in 2019. I analyze the discursive strategies used in the guides to shift ideas about inherent empirical food truths and to map how Canadian Food Guides contribute to the national governance on nutrition as well as the widespread individualization of healthy eating through nutricentric and neoliberal governing strategies. While I focus on Canadian Food Guides, my analysis will resonate with other national eating guidelines, given the overlap in economic and political structures among Western countries and the symmetries in many of the guides themselves.

Amidst widespread concern over the association between food choice and chronic disease, Canada's Food Guide (like other international guides) attempts to alter population eating habits without directly intervening in food industries and placing restrictions and regulations on ingredients, processing methods, and/or advertising. Like its cousin the food label, national Food Guides occupy a contentious space between the increasingly aggressive tactics of food industry and the increasingly urgent consumer need for clear nutritional advice (Mayes, 2014). Its role has long been controversial and the current edition is no exception. As my school-aged self can attest, Food Guides (like food labels) are not simply informational documents but have commonly been used as educational, and thus propaganda, tools (Frohlich, 2017). Beyond the seemingly simplistic messages, the Food Guide frames core categories of food, indicates best-practice habits, and upholds exceedingly individualized responsibilities for health and has done so since its inauguration in 1942.

A brief history of the guide, 1942–2007

Written against the backdrop of national food rationing of the 1930s, as well as Second World War military efforts, the 1942 "official" Food Rules

CANADA'S OFFICIAL FOOD RULES

These are the Health-Protective Foods
Be sure you eat them every day in at least these amounts.
(Use more if you can)

MILK—Adults—½ pint. Children—more than 1 pint. And some CHEESE, as available.

FRUITS—One serving of tomatoes daily, or of a citrus fruit, or of tomato or citrus fruit juices, and one serving of other fruits, fresh, canned or dried.

VEGETABLES (In addition to potatoes of which you need one serving daily)—Two servings daily of vegetables, preferably leafy green, or yellow, and frequently raw.

CEREALS AND BREAD—One serving of a whole-grain cereal and 4 to 6 slices of Canada Approved Bread, brown or white.

MEAT, FISH, etc.—One serving a day of meat, fish, or meat substitutes. Liver, heart or kidney once a week.

EGGS—At least 3 or 4 eggs weekly.

Eat these foods first, then add these and other foods you wish.

Some source of Vitamin D such as fish liver oils, is essential for children, and may be advisable for adults.

Figure 2.1 Canada's official food rules, 1942.
Source: Health Canada. (2007). *History of Canada's Food Guide from 1942 to 2007* (Catalogue number H164-244/2019). Retrieved from https://www.canada.ca/en/health-canada/services/canada-food-guide/about/history-food-guide.html

were established as a strategy quite literally to strengthen the nation (see Figure 2.1). Developed by the federal government in collaboration with the Canadian Council on Nutrition, the inaugural edition responded to the generally poor access to food, lack of money for food, and post-Depression malnutrition among Canadians (Health Canada, 2007c). Ian Mosby (2014) documents that as many as 40% of Canada's WWII military recruits were denied due to inadequate nutrition, rendering the home front weakened by malnourished soldiers. Summed up by the headline of a 1942 *Saturday Night* magazine, "Canada's faulty diet [was] Adolf Hitler's ally" (quoted in Mosby, 2014, p. 51). Through the creation of Canada's first Food Rules, eating well was framed as a patriotic duty. The first Food Rules outlined six major food groups—milk, fruit, vegetables, cereals and breads, meat and fish, and eggs—and featured specific recommendations such as the consumption of one serving of tomatoes, citrus fruit or its juices, and one serving of potatoes daily, as well as one serving of liver, heart, or kidney meats weekly. This inaugural Food Rules and its compendium materials, such as score sheets for daily meals, lesson plans for educators, shopping lists for consumers, and other leaflets, were circulated via the radio, press releases, magazine articles, and newspapers. A series of leaflets produced under the title "Check Your War Efficiency" were also inserted into weekly pay envelopes, covering topics such as breakfast, lunch, and the role of milk in healthy eating (Health Canada, 2007c).

Relatively minor changes were made to Canada's Food Rules in 1944, including removing the term "official" from its designation. Other changes included the recommendation of slightly greater quantities of milk, because of low riboflavin levels in the population, and the removal of kidney and heart meats due to their scarcity of supply, especially for mass consumption. The recommendations for liver remained because of its "distinct nutritional characteristics" (Health Canada, 2007c). Cheese and eggs, due to their protein content, were collapsed into the meat and fish category, and with that, the total number of food groups in the 1944 rules was reduced to five. Earlier marketing efforts of the food rules were curtailed because of cutbacks on federal spending. As such, much of the education efforts for the 1944 rules were centred around working with provinces and communities. For the first time, Canada's Food Rules featured pictures (militarized images in relation to the wartime ideologies outlined above) alongside the major food groups. Earlier compendium materials such as the shopping list were updated and broadened, highlighting the represented food groups. A fact sheet devoted to food budgeting was added, as well as a sheet on how to avoid excess intake. In addition to the patriotic messaging associated with the inaugural edition, the 1944 Food Rules underscored the awareness of famine in other parts of the world. Canadians were urged to "do [their] bit for hungry humanity by conserving food. Buy less. Use less. Waste nothing" (Nutrition Division, 1946, quoted in Health Canada, 2007c). Canadians were encouraged to eat well to strengthen the home front and to mitigate world hunger.

The five food groups remained in the 1949 version of the Food Rules, but several other relatively minor edits were made (see Figure 2.2). The recommendation of "at least" was added to the milk category to accommodate the possibility for greater energy needs by some users of the guide. Bread was no longer limited to "Canada Approved Vitamin B bread", and references to butter now included fortified margarine, thus initiating a 50-plus-year debate on saturated fats (Taubes, 2008). Previous advice about fish oil was replaced with a recommendation for vitamin D, responding to a deficiency among Canadian children and reflecting the wider "vitamania" in nutrition research at the time (Apple, 1996). Similar to its predecessor, a plea to avoid excess intakes crept into compendium materials. The limiting of excess was partly due to food scarcity in other parts of the world, but also a shift to the "eat less" messages of future editions of the guide.

Between 1949 and 1961, when the next iteration of the guide was published, Canadians witnessed major changes in the methods of food processing, storage, and transportation, and correspondingly experienced a wider availability of food throughout the year. Amidst growing industrialization, a contradictory pattern emerged: as "food choices broadened [the] language [around food advice] softened" (Health Canada, 2007c). For the first time, Canada's Food Rules became Canada's Food *Guide*. Correspondingly, the

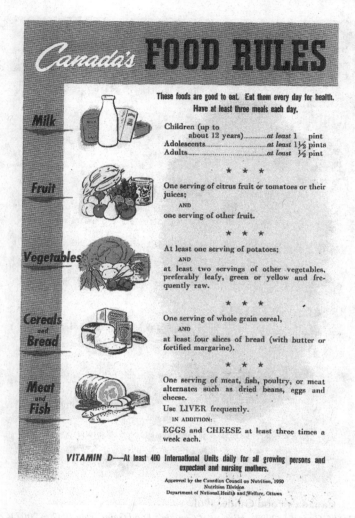

Figure 2.2 Canada's food rules, 1949.

Source: Health Canada. (2007). *History of Canada's Food Guide from 1942 to 2007* (Catalogue number H164-244/2019). Retrieved from https://www.canada.ca/en/health-canada/services/canada-food-guide/about/history-food-guide.html

1961 guide (akin to many of its descendants) stressed flexibility in healthy eating, recognizing that many different dietary patterns could satisfy nutrient needs, and setting up a broader shift towards individualized choice. The other major change in the 1961 guide is that it underwent a colourful facelift, moving away from the predominantly grey, text-based look of previous editions. As depicted in Figure 2.3, the five food groups were now arranged into colour-coded horizontal bars, each clearly differentiating a separate food

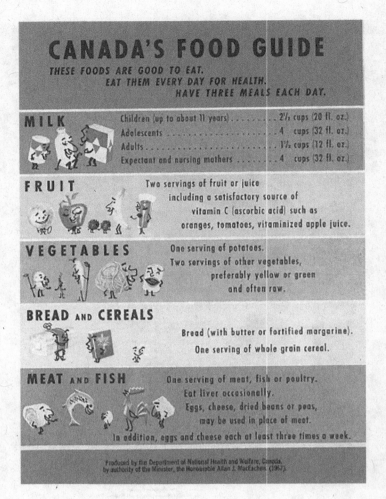

Figure 2.3 Canada's Food Guide, 1961.
Source: Health Canada. (2007). *History of Canada's Food Guide from 1942 to 2007* (Catalogue number H164-244/2019). Retrieved from https://www.canada.ca/en/health-canada/services/canada-food-guide/about/history-food-guide.html

group. While food group classification is commonly used in many international models, critics point out that it falsely represents the idea that certain nutrients can only be found in certain food groups, such as calcium in dairy and iron in meat (Painter & Rah, 2002). In actuality, there is much overlap between food groups. Case in point is that despite debate about collapsing fruits and vegetables into a single group in the 1961 revisions, they remained separate for purposes of teaching and familiarity (Health Canada, 2007c). The newly colour-coded, more image-based guide was perhaps an attempt to communicate increasingly complex nutritional guidance more efficiently, but it also had the added advantage of benefitting the food industry.

Fast-forward 16 years to the 1977 edition, which underwent more significant visual changes. Once entirely text-based, the guide now featured a front page almost exclusively of images. The horizontal bars of the previous edition were replaced with a circular, wheel-like graphic around a caricature of the sun (see Figure 2.4). Each food group's piece of the circular pie was equal, despite the differentiation in portion sizes, listed outside the graph.

Figure 2.4 Canada's Food Guide, 1977.

Source: Health Canada. (2007). *History of Canada's Food Guide from 1942 to 2007* (Catalogue number H164-244/2019). Retrieved from https://www.canada.ca/en/health-canada/services/canada-food-guide/about/history-food-guide.html

The reverse page of the guide featured more text-based information, at least for those who sought it out. In addition to the changes in visual design, the 1977 guide included 30 other content-based alterations (Health Canada, 2007c). Given their nutritional overlaps, fruits and vegetables were combined into a single food group. Ranges were added to the serving suggestions (e.g., four to five servings of fruits and vegetables, as opposed to the previously prescribed two servings), which promoted more flexibility for users as well as a jump in the number of recommended servings. Metric units were used in place of the imperial measurements of past guides. The dairy category was renamed to "milk and milk products", opening the door for other dairy items such as cheese and yogurt. "Meat and alternatives" replaced meat and fish. And a controversial statement that "enriched" grain products could be used in place of whole grain enabled boxed cereals to be included as a government-endorsed food item, a major coup for the cereal industry.

The 1982 edition of the guide featured minor changes, especially compared to the significant rewrite of the previous guide. Revisions in this edition were promoted by increasing evidence on the relationship between diet and cardiovascular disease (Health Canada, 2007c). As depicted on page 2 of the guide, the emphasis on "variety" continued, but it was supplemented with a message of "energy balance" to stress energy intake with energy output, as well as new official messages on "moderation" that advised limited amounts of fat, sugar, salt, and alcohol. At first glance, these changes may have appeared minor, but they signified a major shift in dietary advice in Canada. The initial goal of preventing nutrient deficiencies was now replaced by the aim of reducing chronic diseases through individual diet choices. The "eat more", strengthening strategies of early food rules had shifted to the "eat less" messages of the burgeoning neoliberal era, where citizens were asked to make healthful food choices in the face of industry influence and increased food marketing, growing changes to food production and quality, and, confusingly, increased portion recommendations by the guide itself. The era of what historian Harvey Levenstein (2003) calls "negative nutrition" (i.e., the focus on eating less and avoiding "bad" calories) had arrived and would only amplify in subsequent years.

The 1992 edition marked another change in title, image, and messaging for the guide. Further entrenching a neoliberal rhetoric of food choice, in Canada's Food Guide to Healthy Eating the word "choose" appeared five times on the front page alone (see Figure 2.5). The heightened focus on choice was embroiled into the shift from a "foundation diet" that promoted a specific, minimum intake of food and nutrient consumption, to a "total diet" approach that expanded overall food options (Bragg & Nestle, 2017). While previous Food Guides listed specific intakes (such as consume potatoes daily) alongside narrower intake ranges, the 1992 guide hiked the maximum serving sizes to 10 servings of fruits and vegetables, 12 servings of grains, 4 servings of dairy, and 3 servings of meat and alternatives, *per day*. The 1992 serving sizes doubled and in some cases tripled earlier dietary recommendations, all while donning a language of choice and moderation

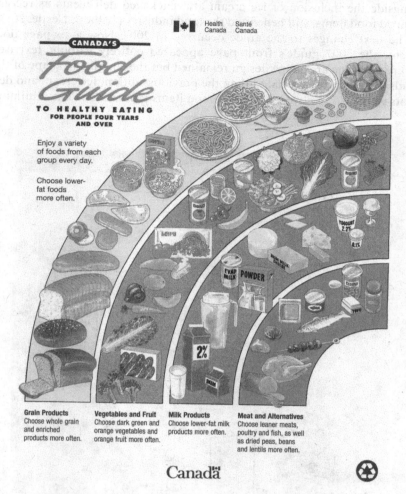

Health   Santé
Canada   Canada

CANADA'S

Food Guide

TO HEALTHY EATING
FOR PEOPLE FOUR YEARS
AND OVER

Enjoy a variety
of foods from each
group every day.

Choose lower-
fat foods
more often.

Grain Products
Choose whole grain
and enriched
products more often.

Vegetables and Fruit
Choose dark green and
orange vegetables and
orange fruit more often.

Milk Products
Choose lower-fat milk
products more often.

Meat and Alternatives
Choose leaner meats,
poultry and fish, as well
as dried peas, beans
and lentils more often.

Canadä

Figure 2.5 Canada's Food Guide, 1992.
Source: Health Canada. (2007). *History of Canada's Food Guide from 1942 to 2007* (Catalogue number H164-244/2019). Retrieved from https://www.canada.ca/en/health-canada/services/canada-food-guide/about/history-food-guide.html

(Falbe & Nestle, 2013). These hikes were accompanied with a short note on activity levels and varying energy needs at the bottom of page 2. The front-page graphic was changed to a rainbow in an attempt to display the four food groups proportionally.[1] Grain products are featured at the top of the arc (i.e., with the largest number of recommended servings), followed by vegetables and fruit, milk products, and meat and alternatives at the bottom of the arc (i.e., with the smallest number of recommended servings). In comparison to past visual representations, dairy and meat products are minimally displayed. However, the overall hike in daily recommended servings, alongside the inclusion of ice cream and packaged deli meats as recommended food items, still benefitted the food industry (Falbe & Nestle, 2013).

The next changes to the guide occurred in 2007. Now a six-page document, the 2007 guide's front page appeared entirely without text (see Figure 2.6). The arc-shape design remained but the length and shape of the bands were less differentiated than the previous edition. Ice cream and deli meats were also removed as specific food items represented. The minimum

Figure 2.6 Eating well with Canada's Food Guide, 2007.

Source: Health Canada. (2007). *Eating well with Canada's Food Guide* (Catalogue number H164-38/1–2007E). Retrieved from https://www.canada.ca/en/health-canada/services/canada-food-guide/about/history-food-guide/eating-well-with-canada-food-guide-2007.html

serving sizes were further increased in two categories (fruits and vegetables, and grain) and maximum serving sizes were decreased in two categories (grain and dairy), alongside a deepened focus on nutrients, disease reduction, and health. Page 3 of the guide featured examples of serving sizes, with a brief discussion on the role of oils and fats. Relying on the ever-insidious language of choice (and derivatives thereof), page 4 outlined a dozen or so food instructions, including statements such as "select lower fat milk alternatives", or "have vegetables and fruit more often than juice" (despite juice being represented as a serving of fruit on the previous page). Perhaps the biggest change with the 2007 edition was the attempt to individualize food advice. Pages 2 and 5 of the Guide differentiated portion sizes based on sex and age; that is, there were portion sizes for children, teenagers, and adults, and the adult category was further subdivided between under 50 and over 50. The last page of the guide offered a hodgepodge of advice on being active, limiting trans fats, reading food labels, and eating well, alongside a paradoxical statement "to savour every bite" (Health Canada, 2007b). The 2007 Food Guide was also launched concurrently with a suite of additional resources, including an interactive website, the My Food Guide mobile application, and a 50-page resource for educators (Health Canada, 2016).

Also in 2007, Health Canada released Eating Well with Canada's Food Guide: First Nations, Inuit and Métis, henceforth the "FNIM Food Guide" (Health Canada, 2007d). Building on the 1988 Northwest Territories Food Guide (Northwest Territories Health and Social Services, 2001) and the 2005 Nunavut Food Guide (Nunavut Department of Health and Social Services, 2005), which were territory-specific documents, the FNIM Food Guide claims to reflect the values, traditions, and food choices of First Nations, Inuit, and Métis populations while also delivering nutrition-based advice to Indigenous communities across Canada (Health Canada, 2010). The front page of the guide features a circle (a sacred symbol for Indigenous groups) with the food items on the outer ring and images of traditional hunting and trapping practices at the centre. The FNIM guide references traditional foods such as game meats and bannock,[2] alongside store-bought Westernized foods such as cheese and broccoli. There are other notable differences between the two 2007 guides. The FNIM guide is half the length of the non-Indigenous guide, features no discussion of deciphering food labels, and with the exception of dairy, lumps teens and adults into the same age category. The guide was translated into Inuktitut, Ojibway, Plains Cree, and Woods Cree, which, as the Government of Canada (Health Canada, 2010) outlines, are the most commonly spoken Indigenous languages in Canada, despite the aim of the FNIM guide being to offer nutritional guidance to Indigenous communities across the country.

Critiques of Canada's Food Guides

Since their inauguration in 1942, Canada's food rules and guides, including the FNIM guides, have been controversial. The attempt to manage population health through dietetic education has generated heated debate about what is (and is not) represented. From free-market enthusiasts who resist any attempt to regulate food knowledge, to public health activists who rail against industry influence, the critiques of Canada's Food Guides are lengthy. Across disciplinary divides, dietitians, doctors, journalists, and public health advocates agree that Canada's Food Guide is inadequate at best and deeply problematic at worst. Ann Hui (2017) deems the 2007 guide a veritable mess; Marni Soupcoff (2017) contends that the guide "has been dominated by shoddy science and short-sighted advice for decades"; and in an open letter to Health Canada, a group of Canadian Physicians and Allied Health Care Professionals (2016) call out the guide for contributing to the problem of childhood "obesity".[3] I outline critiques of the guide to situate a deeper understanding of its role in creating and normalizing individualized and neoliberal frameworks of healthy eating. These critiques also contextualize some of the changes to the long-awaited 2019 version of food-based dietary guidelines in Canada.

Canada's Food Guides have long been influenced by food industries, closely co-branded with the interests of Canada's main domestic products: dairy, wheat, and meat (Mosby, 2014). Between one-quarter and one-third of the advisory committee that creates the guide is directly representative of industry, including Food and Consumer Products of Canada, which represents food giants such as Kraft, Dole, Hershey, and Kellogg's (Hyslop, 2014). There are a number of examples where Food Guide recommendations are concomitant with the interests of industry. As Katie Hyslop (2014) points out, with the exclusion of one reference on the last page of the 2007 guide that encourages consumers to "limit their intake of foods and beverages high in calories, fats, sugar or salt", the 2007 manual was otherwise wholly silent about fats and highly processed foods. Further, health claims associated with the 2007 guide, such as choosing products "low in fat" or those with "no added sugar" mirror many of the marketing claims used in processed foods such as cereals, granola bars, and other snack products. Moreover, the 2007 recommendations stated that half of the six to eight servings of grains should be whole grain, which as Sophia Harris (2016) adds, left another (notably generous) half to refined grain products, including the highly processed cereals featured prominently in the 2007 edition.

Despite the Food Guide circulating as a tool to help mitigate "obesity", it has been accused of contributing to obesogenic environments (Jessri & L'Abbé, 2015; Kondro, 2006). Specifically, the guide has been assessed by food experts as being too lenient on the intake of sugar (including juice), salt, refined grains, starchy vegetables, red meats, and trans fats—food items with suspected links to weight-gain (Hui, 2017). Another widespread critique linked to the fear of high "obesity" rates is the curiously high daily serving recommendations, which spiked in the 1992 edition. In 1992, the maximum recommended daily servings included 12 servings of grain, ten of fruits and vegetables, four of dairy, and three of meat. In 2007, the high end of daily recommendations lowered slightly to ten servings of fruits and vegetables, eight servings of grain, and three servings each of dairy and meat. When translated into calories, these serving recommendations equalled about 4,000 calories, which is roughly *double* the recommended daily intake (Bragg & Nestle, 2017). Like its US equivalent, the 2007 Canadian guide failed to distinguish between good and bad fats and between good and bad carbohydrates, insufficiently warning consumers about the health risks of overconsumption of refined or highly processed grains and of saturated and trans fats (Taubes, 2008).

Another way in which the 2007 guide failed the mandate of public health was the confusing nature of the document itself. Rather than guiding people to nutritional clarity, it has been criticized as adding to the growing confusion around dietary advice. What began as a one-page leaflet grew into a six-page document, replete with information on aging, activity levels, and food labels as well as addendum materials, interactive tools, and mobile phone applications.[4] For many consumers, the 2007 Food Guide was "long, cumbersome, and not the best tool to put in your back pocket" (Sue Mah, quoted in Hui, 2017). Aside from the growing length of the guide, critics also point to serving-size measurements that are impractical and confusing (Loney, 2015). Originally measured in pints, serving sizes over the years have alternatively been measured in ounces, grams, millilitres, teaspoons, tablespoons, and/or cups, confusing to many who may not know conversations rates or comprehend empirical quantities. Serving sizes, moreover, are not standardized, so one item might be listed in cups, while another in ounces (Yin Man Chan, Scourboutakos, & L'Abbé, 2017). As Scrinis and Parker (2016) assert, empirical food language requires a certain degree of scientific literacy, which they add, is least likely to be present in lower-income groups due to lower levels of formal education; paradoxically, these groups are most likely to be consuming diets poorest in nutritional quality. Published Food Guides, alongside food labels, "popularize and socialize a discourse that had previously been reserved for scientists" (Mudry, 2009, p. 47), which adds to a growing confusion in diet- and health-related advice directed at consumers.

Also contributing to this confusion and pointing to the selective representation of nutritional knowledge is the classification of specific food items. The 1942 Food Rules first outlined six major food groups, which moved to five in 1944, and in 1977 dropped to four. In the US, the number of food groupings was as high as 12 in 1933, eventually dropping to four in 1956 (Mudry, 2009). While the four food groups of grains, vegetables and fruits, dairy, and meat may seem commonplace in North American contexts, there are notable cultural differences of food classification. In Germany and Sweden, for instance, root vegetables are listed as the base food group instead of grain products. In some countries, potatoes are included in the vegetable category and in others they are classified in grains. Beans and legumes are also varyingly listed between meat and vegetable groupings.

Dairy is not included as a food item in the Philippine Food Guide, and in the Mexican guide it is grouped with other foods of animal origin (Painter & Rah, 2002). Skeptics of the current Canadian Food Guide classification have also questioned why dairy is its own category when all of the nutrients found in it can also be found in other food items (Harris, 2016). Moreover, milk and alternatives are listed in the Canadian Food Guide as good sources of calcium, but not as good sources of protein, while alternative sources of calcium (such as those found in vegetables) are completely ignored. The very designation of food groups and their perceived health benefits is a cultural and discursive practice.

Canada's Food Guide and its dietary advice have also been criticized as Eurocentric and assimilationist (Mosby, 2014). The reliance on and representation of European food items as well as the steadfast focus on dairy has had little application to East Asian, Indigenous, and other ethnic Canadian groups who typically subscribe to different dietary patterns and who do not use dairy in traditional cuisine. Akin to the US context, dominant dietary wisdom "discourage[d] new immigrants from continuing their ethnic food traditions and encourage[d] [...] whitening their meals with cream sauces and fresh milk as a beverage on the side" (DuPuis, 2015, p. 91). While the 2007 guide makes some effort to include "ethnic" food items such as couscous, tofu, and *kefir*, these items are more tokenistic than a genuine attempt to overhaul the Eurocentric advice of the guide. The 2007 Food Guide fails to offer a wider range of food information that would meet the nutritional needs of a diverse Canadian population and accurately reflect the way that many Canadians eat (Group of Concerned Canadian Physicians and Allied Health Care Providers, 2016; Hui, 2017). The ethnocentrism evident in the history of the guide further others non-European Canadians by stigmatizing the food practices and norms of non-European groups (Arceño, 2016). By virtue of the foods and dietary patterns shown (and not shown), Canadian Food Guides privilege European (and by extension white) food habits as superior to other traditional cultural models of healthy eating. Despite what its name might represent, Canada's FNIM Food Guide is also painfully Euro- and ethnocentric.

"Racial indigestion" and Canada's FNIM Food Guide

Kyla Wazana Tompkins' (2012) coins the concept of racial indigestion to examine the social, symbolic, and material practices through which colonial practices produce and maintain racial (and other systemic) food and social inequalities. While Tompkins' analysis focuses on the antebellum period of the late 18th and early 19th centuries in the United States, I offer an application of her concept of racial indigestion to the ongoing colonial politics of the Canadian Food Guides as they govern and regulate FNIM food systems and eating practices. I find her theorizing of racial indigestion particularly useful for reflecting and articulating the dual processes and effects of literal indigestion that arises from Indigenous food insecurity, alongside the symbolic unease and malaise of contemporary colonial relations in present-day settler Canada.

Aside from tokenistic, superficial differences, the FNIM Food Guide mirrors the advice of its non-Indigenous counterpart. While the cover page centralizes Indigenous imagery, the outer ring of the guide's first page, akin to the whole second page, is largely white-washed. What's shown, with the exception of games meats, a small piece of bannock, a few berries (alongside packaged ones), and a short, written reference to wild plants, are store-bought southern foods including juice, fresh and frozen southern vegetables, packaged cereals, yogurt, milk, and canned fish—the very same food items that have been subsidized in the last 50 years of government food programs in northern and Indigenous communities (Burnett, Hay, & Chambers, 2016). Dairy features prominently in the FNIM guide despite the fact that dairy is not a traditional food item for Indigenous communities and that most Indigenous Peoples are lactose intolerant (Bhatnagar & Aggarwal, 2007). The guide also presents "Indigenous" food advice without any recognition of Indigenous food knowledge. In the words of Anishinaabe food activist Winona LaDuke (2015), "For us, food comes from our relatives, whether they have wings or fins or routes, that is how we consider food. Food has a culture. It has a history. It has a story. It has relationships". The relationality of Indigenous food knowledge and its links to the land, community, spirit, and history are erased through assimilationist strategies that encourage Indigenous Peoples to eat in the same way settlers do.

In addition to reproducing white, European ideologies of food and healthy eating, the FNIM guide has little practical application to the lived realities of many Indigenous people. Because of centuries of colonial dispossession, generations of government policies have limited and undermined Indigenous food security and food sovereignty. Laws restrict the types of animals that can be hunted, the times of year that hunting can take place, and the number of animals a hunter can kill and keep, all of which affect traditional food gathering practices (Martin, Jamal, Ramsay, & Stohart, 2016). Hunting, trapping, and gathering have further been affected in Indigenous communities by the depletion of wild food sources as a result of mining, damming, and development projects; treaty-imposed settlements; and the residential school system, which has disrupted the intergenerational transmission of land-based skills (Haman, Fontaine-Bisson, Pilon, Lamarche, & Robidoux, 2017). Even the reliance on store-bought foods featured in the FNIM guide sidesteps the many economic barriers of purchasing food in many remote and northern communities. On average, foods purchased in the north are two to three times more expensive than the same foods purchased in the south (Martin et al., 2016). Part of this inflation has to do with extra transportation and shipping costs. However, as LeBlanc and Burnett (2017) point out, the North West Company's stores, which hold a near monopoly in the North, recorded $143.3-million profits in 2012. Between the accumulated effects of government-imposed regulations, environmental degradation, the broader loss of traditional culture from colonial and racist policies, and a lack of affordable market-based options, fast food chains are often one of the only food "choices" that remain for many Indigenous Peoples in northern and remote communities (LeBlanc & Burnett, 2017).

The specific messaging directed to Indigenous communities in the FNIM Food Guide is also noteworthy. Outlined on page 3 of the guide are two prominent textboxes. The first reads, "Respect your body...your choices matter" (Health Canada, 2010) and encourages readers to limit junk food items such as pop, sports drinks, and chips. While this advice appears in the non-Indigenous guide, it is buried on the last page and takes up significantly less textual and visual space. For me, the words "respect" and "choice" negate the severe lack of access to drinkable water in many FNIM communities across Canada and to affordable, fresh food as detailed above. As of 2018, 177 Indigenous communities across Canada do not have access to fresh drinking water and many of these have been on boil water advisories for a decade or more (Council of Canadians, 2018). Where fresh drinking water is not available, pop and sports drinks (the very drinks Indigenous communities are encouraged to avoid) are often the most available, affordable "choices". The second box in the FNIM Guide reads, "People who do not eat or drink milk products must plan carefully to make sure they *get enough* nutrients" (Health Canada, 2010, emphasis added). The phrasing "get enough" mirrors the Dairy Farmers of Canada's 2012 ad campaign, which aimed to simplistically equate dairy consumption with overall good health (Overend, 2016). As Haman et al. (2017) confirm, traditional Indigenous diets, which provided essential fatty acids, vitamins, and minerals did not rely on dairy as a food source. Most traditional diets were carnivorous, with low amounts of carbohydrates and sugars, which were obtained seasonally from plants and berries. While the box briefly mentions wild plants, bannock, fish bones, shellfish, nuts, and beans as traditional dairy-equivalent food sources, these items do not contradict the prominence of dairy featured throughout the guide—a food item that causes indigestion for Indigenous Peoples, both at the literal level of digestibility and at the symbolic level of continued colonial foodways undermining Indigenous food security and sovereignty.

Tompkins' (2012) concept of racial indigestion "look[s] beyond food itself to consider practices and representations of ingestion and edibility [...] in which objects, people, and political events are metaphorically [...] figured through the symbolic process of eating" (p. 2). In applying this to Food Guide messaging directed at FNIM populations, the question becomes not what *is* healthy, but who gets to eat well and live well within settler-colonial, neoliberal formulations of health? By moving the discussion beyond the "what" of food choice towards the "how" and "why" of food availability and affordability pushes "towards a critique of the political beliefs and structures that underlie eating as a social practice" (Tompkins, 2012, p. 2). Food

Guides and labels are part of the larger epidemic of neoliberal approaches to health, food, and nutrition—approaches that place the responsibly for healthy eating outside the social, structural, and political conditions that create food inequality, which, unfortunately, were only amplified in the 2019 edition.

"Eat well. Live well": Canada's Food Guide, 2019

The long-awaited 2019 Food Guide (see Figure 2.7) marks another era in Canadian Food Guide advice. Released on January 22, 2019, 12 years since the previous edition, the new Food Guide makes significant changes to the

Figure 2.7 Canada's Food Guide, 2019.
Source: Health Canada. (2019). *Canada's Food Guide*. Retrieved from https://food-guide.canada.ca/en/

food groups, images, and advice. Taking a radical step away from the iconic four food groups used for the past four decades, the new Food Guide classifies three food groups: "vegetables and fruit", "protein foods", and "whole grains". Health Canada delisted dairy and meat as distinct groups altogether, collapsing them under protein foods. Even under the category of protein foods, dairy and meat are de-emphasized with the specific recommendation to "choose protein foods that come from plants more often" alongside the visual images of tofu, beans, and legumes (Health Canada, 2019). Moving away from the rainbow arc used since 1992, the new Food Guide adopts the image of a plate, with the corresponding advice that half one's plate should be vegetables and fruit, one-quarter protein foods, and one-quarter whole grain foods (the word "whole" added to the grain category). Specific portion sizes and recommended daily intakes used in the entirety of the Food Guide's history are also removed, in favour of proportions. Water is listed as the recommended drink of choice, especially in place of sugary drinks, including juice, which had featured prominently in previous editions. The other major change to the advice is the addition of more holistic dietary recommendations on page 2 of the guide. This advice includes: "be mindful of your eating habits", "cook more often", "enjoy your food", "eat meals with others", "use food labels", "limit foods high in sodium, sugars or saturated fats", and "be aware of food marketing" (Health Canada, 2019). Health Canada (2019) undertook years of public and private consultations with thousands of stakeholders, including dietitians, food scientists, academics, medical doctors, and the general public. Since its publication, the Guide has been translated in 28 languages, including nine Indigenous languages.

As with previous editions, the most recent version is also not without controversy. Many health professionals and other stakeholders applaud the guide for its strong stance against industry. Responding to critiques that the previous guide was too heavily influenced by commercial food industries, the new Guide uses only "the best available science", which meant excluding industry-funded research on the basis of a conflict of interest (Health Canada, 2019). The Guide's obvious shift towards plant-based foods is also seen as an affront to the Canadian meat and dairy industries, which have long held conspicuous places within it. With tensions between Health Canada and Agriculture Canada, the new Guide has been praised for finally putting the health of Canadians before the needs of industry (Young, 2019) and has been hailed as a major coup "for evidence-based policy" (Shelley, 2019). In particular, the guide's focus on plant-based foods is a step towards mitigating rising cardiovascular disease, colon cancer, and type 2 diabetes among Canadians (Hui, 2017). The 2019 edition has also been praised for its simplicity, moving away from the overly confusing guides of years past, and for being more in line with other countries' food advice—namely, the much-lauded Brazilian Food Guide, which focuses on plant-based eating and more holistic food advice (Tsui, 2019). The shift to include not just "the what" of food but also "the how", with advice such as cook more often, read food labels, and eat with others, is a welcome addition for many supporters of the new Guide, helping Canadians develop more critical practices towards food and eating.

Unsurprisingly, the Canadian beef and dairy industries were both particularly outspoken against the new Guide, claiming it has an "environmentalist agenda" (Kirkey, 2019) and denouncing its misguided, dangerous advice that presents animal- and plant-based proteins as equivalent. Danielle Smith (2019), a writer for *Calgary Herald*, reminds her readers that meat, eggs, and dairy are all complete proteins and that vegans have to work harder to get a full complement of amino acids. Lucie Edwardson (2019), quoting the policy manager for the Alberta Beef Producers, states that beef is a "nutrient-dense food" that should not be overlooked by the advice of the 2019 edition. While some applaud the simplicity of the new Guide, others criticize it as being too vague in its move away from specific portion sizes, daily-recommended intakes, and for being out of touch with the way most Canadians eat. As Michael Graydon—a chief executive at Food and Consumer Products of Canada—explains, the day-to-day lives of many Canadians do not enable home-cooked meals with family and friends: "During the week, people, they're pushed. Their kids are in activities. People are working. It's hard" (quoted in Edmiston, 2019). Others have criticized the Guide as unrealistic because of how expensive fresh fruits and vegetables are, especially during long Canadian winters. While beans and legumes are typically much cheaper than meat and dairy, the focus on eating and preparing fresh fruits and vegetables would be difficult for many Canadians to sustain economically and timewise (Picard, 2019; Tsui, 2019).

While I think the 2019 edition does a good job at attempting to move away from an overly quantified approach to food towards one that encourages Canadians to also think about the context of their eating habits, Food Guides, like food labels, still place the onus for healthy eating too squarely on the individual, as opposed to the conditions that produce food inequality. Underscored in the overview of the history of Food Guides in Canada (as well as international models) is the degree to which empirical, scientific, and nutricentric approaches to food are prioritized over cultural, relational, ecological, environmental, and/or structural understandings of food access. Food Guide and food label messaging draws overwhelmingly on a language of quantification when making claims concerning healthy eating. From serving sizes to percentage of daily intakes, "there has been a shift from *eating* food to *reading* food" (Frohlich, 2017, p. 165, emphasis in original). Gyrogy Scrinis (2013) similarly contends that in modern scientific approaches to nutrition, "The role of nutrients has often been interpreted outside the context of foods, dietary patterns, and broader social contexts in which they are embedded" (p. 6). Quantified meanings of food, as Xaq Frohlich (2017) also attests, can be "objectified, abstracted, and decontextualized" (p. 150) in ways that relational, contextual, and cultural knowledge cannot, which can be better deployed in neoliberal food governance projects. The 2019 Food Guide, with its focus on relational aspects of food (i.e., eat meals with others, cook more often, be aware of food marketing) aims to balance the overly quantified approach to eating advocated in previous guides. Its move away from specific portions and recommended daily intakes also seeks a more simplistic, proportional understanding of food and eating.

However, at a time when chronic diseases of adulthood are the leading causes of death in affluent societies, including Canada, the focus on asking individuals to eat more plant-based foods dangerously frames structural access to healthy food as an individual failing. As seen in many historical editions of Canada's Food Guide, the eat more (i.e., strengthening) narratives of early scientific wisdom were replaced with the idea that some foods, if consumed in large quantities, can contribute to and cause chronic illness. In shifting the focus onto individual eating habits, systemic and structural accounts of food access, food security, food knowledge, and the time, money, and ability needed to prepare healthy meals are ignored. Such healthist accounts of nutrition (Crawford, 1980) not only depoliticize insidious social inequalities that mitigate access to health, but concomitantly feed market purposes where health becomes yet another commodity to be bought and sold (Guthman, 2011), summed up in the 2019 Food Guide's new dictum that "eating well" equates to "living well". While food and health have been linked since the Hippocratic teachings of humoural medicine, their association in the neoliberal era has been drastically repackaged. What was once a holistic and complex way of living has been reduced to the specific items we put in our grocery carts. The risk with this kind of ideology is that it creates and maintains misguided ideas that health is solely determined by individual food choices: "That with the correct information individuals can become self-governing subjects responsible for their health via consumer practices" (Mayes, 2014, p. 282). Individual health-based messaging, essentialized in Food Guide advice, dangerously obscures the relationship between food and health insofar as it contributes to the understanding of health as the effect of consumer choice rather than as the outcome of systemic and structural marginalization.

Such a framing is especially troubling given the decades of sound research into the social determinants of health, which consistently show that structural factors, including food availability and affordability, have a much greater influence on health than individual choices do (Shamir, 2008). The social determinants of the health point to the systemic layers of social positioning as structural inequality as chief mitigating factors in opportunities for health. To borrow at length from Geoffrey Cannon (2002),

It is absurd to advise mothers that their children should be physically active, when their schools have sold their play and sports. It is futile to advise communities to consume more locally grown vegetables and fruits, when their local shops are filled with energy-dense products made by manufacturers with marketing muscle. It is insulting to advise adults to cook nourishing food for themselves, when they are obliged to work night as well as day to pay the rent. [...] No strategy designed to prevent food-related diseases that focuses on dietary habits or on food consumption, can be effective. In isolation, dietary guidelines are a distraction and part of the problem, not part of the solution. The issue is not what we eat, it is why we eat what we eat.

(p. 502)

At a time when wealth disparities are ballooning, environmental degradations continue, and industrialization threatens food sovereignty and food security, the solution cannot be more individual-based health messaging. While the advice to consume more plant-based foods, eat meals with others, and hydrate with water may help those with relative economic privilege, it does not address the underlying social inequalities responsible for ill health that affect marginalized groups at greater rates.

While Canadians may no longer be encouraged to eat well for military combat, Food Guide messaging remains tied to discourses of strengthening the nation. Revealing insights about the specific workings of Canadian nutritional health discourse, Food Guides uphold broader political and governing projects in ongoing colonial and neoliberal projects, and, in doing so, distract from systemic and structural discussions of health. To this end, the subsequent three case study chapters on dairy, wheat, and meat will analyze the polarization of the health debates that have surrounded these food items in contemporary Western contexts, in order to rethink individual-based dietary discourse. In taking seriously one of the holistic recommendations of the 2019 Guide itself, specifically that "healthy eating is more than the foods you eat", I argue for a healthy eating paradigm that accounts for the most relevant aspects of healthy eating: those that link to social structure, not to individual food choice. Without it, the most relevant aspects of food and eating will continue to be the ones least discussed.

Notes

1 Other common Food Guide shapes used internationally include the pyramid, the pagoda, a circle, and a plate, also aim to represent "proportionality" in the face of increased market choice (Painter & Rah, 2002).
2 A cultural food icon for many Indigenous groups, pre-colonial versions of this bread were made from a wild plant called camas. The wheat-based bannock more common today is a vestige of the Scottish bannock introduced during colonization (CBC Radio, 2016).
3 The dominant framing of obesity as a public health threat has been debunked on a number of levels. LeBesco (2004) argues that the fat body has become a moral symbol of failed and unproductive citizenship rather than a threat to public health. Gard and Wright (2005) similarly contend that any evidence that equates obesity prevalence and disease incidence is, at best, "an indication of a possible link between size and health for a population", rather than an undeniable truth

(pp. 101–102). I refer to "obesity" in this chapter because it emerges as a critique of the Food Guide; however, I do so in quote marks to denote its contentious status as a disease state.

4 A similar trend emerges in American dietary guidelines. Mayes (2014) itemizes that the 1980 Dietary Guidelines for Americans was a 20-page booklet with 7 recommendations, but, by 2010, the booklet grew to 95 pages with 23 recommendations for the general public and an additional 6 for specific populations (p. 380).

References

Apple, R. (1996). *Vitamania: Vitamins in American culture.* New Brunswick, NJ: Rutgers University Press.

Arceño, M. A. (2016). *On consuming and constructing material and symbolic culture: An anthropology of pictorial representations of food-based dietary guidelines (FB-DGs)* (Unpublished MA thesis). Columbus, OH: Ohio State University.

Bhatnagar, S., & Aggarwal, R. (2007). Lactose intolerance. *British Medical Journal, 30*(334), 1331–1332. doi: 10.1136/bmj.39252.524375.80

Biltekoff, C. (2012). Critical nutrition studies. In J. M. Pilcher (Ed.), *The Oxford handbook of food history* (pp. 172–190). Oxford, UK: Oxford University Press.

Bragg, M., & Nestle, M. (2017). The politics of government dietary advice: The influence of big food. In J. Germov & J. Williams (Eds.), *Sociology of food and nutrition: The social appetite* (4th ed., pp. 127–146). Oxford, UK: Oxford University Press.

Burnett, K., Hay, T., & Chambers, L. (2016). Settler colonialism, Indigenous peoples and food: Federal Indian policy and nutrition programs in the Canadian north since 1945. *Journal of Colonialism and Colonial History, 17*(2), doi:10.1353/cch.2016.0030

Cannon, G. (2002). Nutrition: The new world disorder. *Asia Pacific Journal of Clinical Nutrition, 11*(Suppl), S498–S509.

CBC Radio. (2016, January 31). *Bannock: A brief history.* [Audio podcast]. Retrieved from https://www.cbc.ca/radio/unreserved/bannock-wild-meat-and-indigenous-food-sovereignty-1.3424436/bannock-a-brief-history-1.3425549

The Council of Canadians. (2018). *Safe water for First Nations.* Retrieved from https://canadians.org/fn-water

Crawford, R. (1980). Healthism and the medicalization of everyday life. *International Journal of Health Services, 10*(3), 365–388.

DuPuis, E. M. (2007). Angels and vegetables: A brief history of food advice in America. *Gastronomica: The Journal of Food and Culture, 7*(3), 34–44.

DuPuis, E. M. (2015). *Dangerous digestion: The politics of American dietary advice.* Berkeley, CA: University of California Press.

Edmiston, J. (2019, January 23). Food Guide out of touch with Canadians, unfairly vilifies processed foods, industry groups say. *The Financial Post.* Retrieved from https://business.financialpost.com/news/retail-marketing/not-fair-or-practical-industry-group-says-new-food-guide-too-harsh-on-processed-foods

Edwardson, L. (2019, January 22). Alberta ranchers fear new food guide will curb Canadians' appetite for beef. *CBC News.* Retrieved from https://www.cbc.ca/news/canada/calgary/ab-ranchers-fear-food-guide-leads-canadians-eat-less-beef-1.4988262

Frohlich, X. (2017). The informational turn in food politics: The US FDA's nutrition label as information infrastructure. *Social Studies of Science, 47*(2), 145–171.

Gard, M., & Wright, J. (2005). *The obesity epidemic: Science, morality, ideology.* New York, NY: Routledge.

Group of Concerned Canadian Physicians and Allied Health Care Providers. (2016, December 8). Open letter. Retrieved from https://www.cattlefeeders.ca/wp-content/uploads/2017/08/open-letter-of-physicians-to-health-canada-on-food-guide.pdf

Guthman, J. (2011). *Weighing in: Obesity, food justice, and the limits of capitalism.* Berkeley, CA: University of California Press.

Haman, F., Fontaine-Bisson, B., Pilon, S., Lamarche, B., & Robidoux, M. A. (2017). Understanding the legacy of colonial contact for a physiological perspective: Nutrition transitions and the rise of dietary disease in Northern Indigenous peoples. In M. A. Robidoux & C. M. Mason (Eds.), *A land not forgotten: Indigenous food security and land-based practices in northern Ontario* (pp. 33–50). Winnipeg, MB: University of Manitoba Press.

Harris, S. (2016, March 22). Health Canada reviewing food guide, critics demand drastic changes now. *CBC News.* Retrieved from http://www.cbc.ca/news/business/health-canada-food-guide-1.3501318

Health Canada. (2007a). *Eating well with Canada's Food Guide* (Catalogue number H164-38/1-2007E). Retrieved from https://www.canada.ca/en/health-canada/services/canada-food-guide/about/history-food-guide/eating-well-with-canada-food-guide-2007.html

Health Canada. (2007b). *Eating well with Canada's Food Guide: A resource for educators and communicators* (Catalogue number H164-38/2-2007E-PDF). Retrieved from http://www.publications.gc.ca/site/eng/300616/publication.html

Health Canada. (2007c). *History of Canada's Food Guide from 1942 to 2007* (Catalogue number H164-244/2019). Retrieved from https://www.canada.ca/en/health-canada/services/canada-food-guide/about/history-food-guide.html

Health Canada. (2007d). *Eating well with Canada's Food Guide: First Nations, Inuit and Métis* (Catalogue number H34-159/2007E). Retrieved from http://www.hc-sc.gc.ca/fn-an/alt_formats/fnihb-dgspni/pdf/pubs/fnim-pnim/2007_fnim-pnim_food-guide-aliment-eng.pdf

Health Canada. (2010). *Eating well with Canada's Food Guide: First Nation, Inuit and Métis* (Catalogue number H34-159/2007E-PDF). Retrieved from https://www.canada.ca/en/health-canada/services/food-nutrition/reports-publications/eating-well-canada-food-guide-first-nations-inuit-metis.html

Health Canada. (2016). *Evidence review for dietary guidance: Summary of results and implications for Canada's Food Guide* (Catalogue number H164-193/2016E-PDF). Retrieved from https://www.canada.ca/en/health-canada/services/publications/food-nutrition/evidence-review-dietary-guidance-summary-results-implications-canada-food-guide.html

Health Canada. (2019). *Canada's Food Guide.* Retrieved from https://food-guide.canada.ca/en/

Hui, A. (2017, July 18). Inside the big revamp of Canada's Food Guide. *The Globe and Mail.* Retrieved from https://www.theglobeandmail.com/news/national/a-taste-of-whats-to-come-inside-the-big-revamp-of-canadas-food-guide/article35728046/

Hyslop, K. (2014, October 20). Is Canada's Food Guide past its best-before date? *The Tyee.* Retrieved from https://thetyee.ca/News/2014/10/20/Canada-Food-Guide/

Jessri, M., & L'Abbé, M. R. (2015). The time for an updated Canadian Food Guide has arrived. *Applied Physiology, Nutrition, and Metabolism, 40*(8), 854–857.

Kirkey, S. (2019, January 22). Got milk? Not so much. Canada's new Food Guide drops milk and alternatives and favours plant-based proteins. *National Post*. Retrieved from https://nationalpost.com/health/health-canada-new-food-guide-2019

Kondro, W. (2006, February 28). Proposed Canada Food Guide called "obesogenic." *Canadian Medical Association Journal, 174*(5), 605–606. doi: 10.1503/cmaj.060039

LeBesco, K. (2004). *Revolting bodies?: The struggle to redefine fat identity*. Amherst, MA: University of Massachusetts Press.

LeBlanc, J., & Burnett, K. (2017). What happened to Indigenous food sovereignty in northern Ontario? Imposed political, economic, socio-ecological, and cultural systems changes. In M. A. Robidoux & C. M. Mason (Eds.), *A land not forgotten: Indigenous food security and land-based practices in northern Ontario* (pp. 16–33). Winnipeg, MB: University of Manitoba Press.

Levenstein, H. A. (2003). *Paradox of plenty: A social history of eating in modern America*. Berkeley, CA: University of California Press.

Loney, H. (2015, June 22). Is Canada's Food Guide unrealistic? *Global News*. Retrieved from https://globalnews.ca/news/2042397/is-canadas-food-guide-unrealistic/

Martin, D., Jamal, A., Ramsay, M., & Stohart, C. (2016). *Paying for nutrition: A report on food costing in the North*. Retrieved from https://foodsecurecanada.org/sites/foodsecurecanada.org/files/201609_paying_for_nutrition_fsc_report_final.pdf

Mayes, C. (2014). Governing through choice: Food labels and the confluence of food industry and public health discourse to create "healthy consumers." *Social Theory & Health, 12*(4), 376–395. doi: 10.1057/sth.2014.12

Mosby, I. (2014). *Food will win the war: The politics, culture, and science of food on Canada's home front*. Vancouver, BC: UBC Press.

Mudry, J. (2009). *Measured meals: Nutrition in America*. Albany, NY: State University of New York Press.

Murphy, S. P., & Barr, S. I. (2007). Food guides reflect similarities and differences in dietary guidance in three countries (Japan, Canada, and the United States). *Nutrition Reviews, 65*(4), 141–148. doi: 10.1301/nr.2007.apr.141-148

Neslte, M. (2013). *Food politics: How the food industry influences nutrition and health*. Berkley, CA: University of California Press.

Northwest Territories [NWT] Health and Social Services [HSS]. (2001). *NWT Food Guide*. Retrieved March 19, 2018, from http://www.hc-sc.gc.ca/fn-an/alt_formats/fnihb-dgspni/pdf/pubs/fnim-pnim/2007_fnim-pnim_food-guide-aliment-eng.pdf

Nunavut Department of Health and Social Services [HSS]. (2005). *Nunavut Food Guide*. Retrieved December 7, 2015, from http://circhob.circumpolarhealth.org/item/nunavut-food-guide/

Overend, A. (2016). Mothering discourse and the marketing of dairy as a cancer-fighting food. In F. Pasche-Guignard & T. Cassidy (Eds.), *Mothers and food: Negotiating foodways from maternal perspectives* (pp. 73–87). Toronto, ON: Demeter Press.

Painter, J., & Rah, J. (2002). Comparison of international food guides pictorial representations. *Journal of the American Dietetic Association, 102*(4), 483–489.

Picard, A. (2019, January 22). Canada's new Food Guide is a good upgrade but skirts around issues of inequality. *The Globe and Mail*. Retrieved from https://www.theglobeandmail.com/canada/article-canadas-new-food-guide-is-a-good-upgrade-but-skirts-around-issues-of/

Scrinis, G. (2013). *Nutritionism: The science and politics of dietary advice*. New York, NY: Columbia University Press.

Scrinis, G., & Parker, C. (2016). Front-of-pack food labeling and the politics of nutritional nudges. *Law & Policy, 38*(3), 234–249. doi: 10.1111/lapo.12058

Shamir, R. (2008). The age of responsibilization: On market-embedded morality. *Economy and Society, 37*(1), 1–19. doi: 10.1080/03085140701760833

Shelley, C. (2019, January 23). Food porn it ain't but new food guide rare win for evidence-based policy. *National Post*. Retrieved from https://nationalpost.com/opinion/chris-selley-new-food-guide-is-a-bit-silly-but-a-rare-win-for-evidence-based-policy

Smith, D. (2019, January 25). If you care about the planet, eat more beef. *Calgary Herald*. Retrieved from https://calgaryherald.com/opinion/columnists/smith-if-you-care-about-the-planet-eat-more-beef

Soupcoff, M. (2017, July 20). Our nutritionally confused government tells us how to host dinner parties. *National Post*. Retrieved from http://nationalpost.com/opinion/marni-soupcoff-new-canada-food-guide-is-light-on-nutrition-heavy-on-spiritual-wellness

Taubes, G. (2008). *Good calories, bad calories: Fats, carbs, and the controversial science of diet and health.* New York, NY: Anchor Books.

Tompkins, K. W. (2012). *Racial indigestion: Eating bodies in the 19th century.* New York, NY: New York University Press.

Tsui, V. (2019, January 23). A dietician's perspective on the new Canada's Food Guide [Blog post]. Retrieved from https://eatnorth.com/vincci-tsui/dietitians-perspective-new-canadas-food-guide

Yin Man Chan, J., Scourboutakos, M. J., & L'Abbé, M. R. (2017). Unregulated serving sizes on the Canadian nutrition facts table: An invitation for manufacturer manipulations. *BCM Public Health, 17*, 418.

Young, L. (2019, January 22). Health experts applaud Canada's new Food Guide, though some questions lack of portion sizes. *Global News*. Retrieved from https://globalnews.ca/news/4875553/canada-new-food-guide-health/

3 Dairy
Beyond "got" and "not" milk

In its material form, milk is an opaque substance that complicates easy boundaries between solids and liquids. According to ancient humoural descriptions, dairy was classified as "cold" and "wet" and should be avoided by "phlegmatic" types as it was thought to generate mucus and slow metabolism (Osborn, 2017). By some contemporary food framings, dairy is still considered inflammatory (Myers, 2013). According to wellness magazine *Healthline* (2017b), those with irritable bowel syndrome should avoid eating dairy; it may also contribute to arthritis pain (Healthline, 2017a); and, Tan (2017) claims, dairy increases and thickens mucus production during a cold or flu. By other contemporary associations, fermented dairy products have anti-inflammatory properties (Bordoni et al., 2017) and are claimed to help in the fight against chronic, non-communicable diseases such as heart disease and cancer (Dairy Farmers of Canada, 2012). Depending on who is consulted, dairy is either a healthful food item on one hand or a dietary risk on the other.

Western symbolic and cultural associations with dairy, and specifically with milk, are equally varied (Brady, Millious, & Ventresca, 2016; Valenze, 2011). Dairy ties both to the feminine maternal and masculine athleticism, it sits at the crossroads between the natural and the technological, and its marketing often draws on playful associations with childhood alongside increasingly resolute nutricentric health claims. While the study of dairy is not new to food studies (see, for instance, Atkins, 2010; DuPuis, 2002; Valenze, 2011; Wiley, 2004, 2011, 2014), it endures because of dairy's longstanding, at times sordid, history in human diets and its current ambivalent status in nutritional health discourses. My interest in dairy, akin to my curiosity about meat and wheat, is that it remains contentious and central to contemporary debates concerning singular food truths: a productive case study with which to decentralize nutricentrisms and to open up plural food truths.

Dairy continues to elicit strong responses on both the "got milk?"[1] and "not milk"[2] ends of the debate, ranging from claims that it is the best source of a range of nutrients, to its status as an over-promoted agricultural product. The longstanding debate between the "lactophobes" and "lactophiles"

(Harris, 1985) remains largely unresolved. Two centuries after the inauguration of mass-produced, industrial milk, the got- and not-milk divides continue to be scientifically and epidemiologically murky (Brady et al., 2016; Gaard, 2013). I detail these debates, relying on already existing historical analyses of milk and dairy products—namely, though not exclusively, E. Melanie DuPuis's (2002) influential *Nature's Perfect Food* and Andrea Wiley's *Cultures of Milk* (2014) and *Re-Imagining Milk* (2011). Through my genealogy of dairy in Western culture, I trace the changing discourses of dairy's health-related claims into a contemporary context and contend, in line with the overarching claims of the book, that there is no singular, inherent truth to dairy's healthfulness and that truth claims of dairy shift alongside changing health issues as well as in relation to changing economic and political contexts. I begin with the discourse of naturalization, which provides one of the longest-standing justifications for dairy's presumed healthfulness.

Discourses of naturalization

With deep and enduring links to mothering, the work of nurturing, and the giving of life, dairy (and specifically milk) has enjoyed—albeit at times tenuously—the honour of nature's perfect food. DuPuis (2002) documents the work of moral reformer Robert Hartley, who in his essays on milk in 1842 was the first to make the public case for milk as a uniquely and naturally "perfect food: containing in perfect measure, all the ingredients to sustain life" (DuPuis, 2002, p. 18). Relying on biblical references to milk and honey (i.e., the only two foods produced by animals that are eaten by other animals), Hartley claimed that milk was both nutritionally complete and universally consumed. William Prout, a 19th-century physician, similarly proclaimed milk as "the most perfect of all elementary aliments" because it contained albumen, oil, and sugar, which are all essential to the production of animal life (quoted in DuPuis, 2002, p. 32). The framing of dairy as part of God's divine nutritional plan echoes back to the 17th-century doctrine of signatures. The paradoxes of cow's dairy as a natural food item for humans, however, are many.

As Wiley (2011) documents, the universal human history of milk drinking is of *human breast milk*, not cow's milk, which is a different mammal with considerably different milk. Mid-19th century marketing efforts claimed cow's milk as a "perfect and natural" substitute for breast milk at a time when mothers were entering the industrial workforce in greater numbers. Yet, later scientific analyses of various animal milks would yield relevant differences between human and cow milk. While all mammalian milks are similarly made up of fats, proteins, and sugars structured to aid growth, development, and immunity in young mammals, mammalian milks are far from equivalent. Marine mammals, for instance, living in cold climates produce milk that is very high in fat to help newborns stay warm. Seal's milk, for example, is more than 50% fat, compared to cow's milk, which is only 4% fat (Wiley, 2011). Cow's milk contains about three times as much protein and four times as much calcium but less lactose and fat than human milk. The higher amounts of protein and calcium in cow's milk enable calves to reach 500 lbs by six months of age. Human babies, in contrast, weigh on average 17 lbs by the same age (Wiley, 2011).

Not only do mammalian milks differ in their specific composition of fats, proteins, and sugars, but their composition also changes over the course of infancy. Mammalian milks produced immediately after birth are extremely rich in antibodies, to boost immunity and protect offspring outside the womb (Wiley, 2011). Not only do these antibodies diminish after the first few weeks of infancy but other changes occur in the milk from birth to the weaning phase. With the exception of the commercial push in favour of formula feeding, upheld by many physicians in the 1950s and again in the 1980s (Fomon, 2001), few have questioned that an infant's consumption of its own mother's breast milk is healthy. Since the mid-1800s, however, over the duration that fresh milk has been in commercial circulation many have raised concerns over human consumption of non-human milks (cow, sheep, goat, and buffalo) and about its consumption past childhood. In their study on the symbolic associations of milk, Wilken and Knudson (2008) show that

> milk is associated with the family, and the relationship between parents and children. Providing milk to someone [...] is widely seen as providing care, especially parental care. Milk is perceived as *the* drink that good parents provide for their children and can thus become understood as an extension of the 'milk tie' between mother and child.
>
> (p. 36, emphasis in original)

While milk and honey may be the only two foods produced by animals specifically as food, they are produced for members of that same animal group—a point often negated in the dairy industry's claims to the naturalness of milk drinking.

In an effort to denaturalize the history of milk consumption, food historians have reminded us that while many global cultures have a long history of *dairy* production and consumption, few cultures have a long history of *milk* consumption. Northern European, Middle Eastern, and Mediterranean cultures have long utilized fermented dairy products such as yogurt, *kefir*, and cheese, which persist as parts of these traditional cuisines (Wiley, 2011). The milk in historical and biblical references is not the milk we know today but rather "clabbered milk"—a sour and fermented yogurt-like beverage (Schmid, 2009). Fermented dairy products contain much less lactose and are therefore easier to digest, but they also spoil less quickly, which would have been of heightened importance before pasteurization and refrigeration technologies were developed. Contrary to the naturalization of milk as nature's perfect food, the production and consumption of fresh milk is actually a recent food phenomenon that emerged alongside urbanization and industrialization. DuPuis (2002) details that "fluid milk drinking was an afterthought [...] from colonial times to the mid-19th century, fresh milk was not a major American beverage" (p. 5). Despite its centrality in grocery store isles, vending machines, and school lunch programs today, fresh milk was historically the least consumed form of dairy. Further tainting its naturalized status, when fresh milk first became a mass-produced commodity, it generated one of the preeminent scandals of modern food production.

Sold in the mid-1800s as "country milk", the first mass-produced, commercially available fresh milk was a far cry from nature's most perfect food. As it was needed in larger quantities to feed growing urban populations and to provide a substitute to breast milk for infants of working mothers, the "swill milk" factories of the mid-19th century were established as an economically advantageous way to deal with the residual mush from distilleries and breweries. This mush (or swill), as DuPuis (2002) describes, "was often poured hot into troughs directly from the brewery to the stable" (p. 18). In addition to the nutrient-deprived swill the cows were fed, stables were unsanitary and overcrowded, and cows "huddled together [...] in the stench of their own excrements" (quoted in DuPuis, 2002, p. 18). The end product was a far cry from what we might imagine "country milk" to be: it was a thin, bluish, and nutrient-poor substance, which had to then be adulterated with plaster and starch to make it look more palatable to consumers. Unsurprisingly, swill milk factories became breeding grounds for bacteria and disease. Infant mortality, disproportionally affecting infants of low-income mothers who were forced to work, spiked. By the 1840s, an infant born in the city had only a 50% chance of living to the age of five and the rate of death in the first year was as high as 20% (DuPuis, 2002). While unsanitary water supplies contributed to cholera (the epidemic that most affected infants at the time), swill milk was an even better medium for the disease. Bacteria not only survived in this substance, the protein and other nutrients provided fuel for its growth. Until widespread sterilization efforts in the 1920s, "milk was one of the major public health issues of the latter 19th and 20th centuries" (Gaard, 2013, p. 596). Despite contemporary connections between milk and life, early industrial milk was linked with disease and death, and for that reason has long been subjected to technological interventions.

Unpasteurized milk can contain dangerous microorganisms such as salmonella, E. coli, and listeria (Health Canada, 2013; U.S. Food and Drug Administration, 2012). These can cause food poisoning and lead to very serious conditions including vomiting and diarrhea (Health Canada, 2013; U.S. Food and Drug Administration, 2012). Health Canada (2013) states these bacteria may also lead to life-threatening kidney failure, miscarriage, and death. Other symptoms of foodborne illness that can result from consuming unpasteurized or raw milk include abdominal pain and flu-like symptoms such as fever, headache, and body ache (U.S. Food and Drug Administration, 2012). To render milk safe for human consumption, the first major changes made to fluid milk were through mass pasteurization of the late 19th century (Wiley, 2007). The process of pasteurization heats the milk to burn off any microbial contaminants found in fresh milk. In doing so, its shelf life is increased, enabling greater efficiency of production and wider commercial distribution (Wiley, 2007). However, as Peter Atkins (2010) contends, pasteurization also has destructive effects on the healthfulness of raw milk. In particular, the heat reduces the number of vitamins present. Vitamins A and D are later refortified into pasteurized and fat-free milks to adjust for the effects of technological intervention and to counteract deficiencies like rickets, which was widespread at the turn of the 20th century when fresh milk became more widely available.

Countering any commercial and commonplace notions of milk's naturalness, the history of cow milk production is a history of technological intervention. Throughout the 19th and 20th centuries, other technological interventions enabled the mass production of more diverse dairy products. Wiley (2007) documents how the invention of the cream separator in 1871 allowed for an efficient separation of the cream (i.e., fat) from the rest of fluid milk and enabled the production of skim milk and other fat-reduced forms that are widespread today. The cream separator became a key technology for milk's role as part of a low-fat diet in the mid-20th century. Paradoxically, at a time when skim and fat-reduced milks were being dubbed as part of the solution to rising rates of chronic illness in North America, the cream separator also provided an opportunity for the dairy industry to expand booming cheese production. As North Americans were drinking lower-fat milks, they were consuming more cheese (Wiley, 2011). Homogenization technologies were also being widely used in the mass manufacturing of ice creams, which dissolves and then redistributes the fat molecules in dairy, enabling a richer, smoother taste. Manufacturers added sweeteners such as chocolate and strawberry syrups to milk to make it more palatable, especially to children.

To counteract the decline of milk consumption in North America, respond to increasing public awareness of lactose intolerance, and match the booming business of non-dairy beverages such as sports drinks, the dairy industry's newest technological intervention is the production of acidophilus-enhanced and lactose-free milks and dairy items (Sethi, Tyagi, & Anurag, 2016). In June 2014 the European Union ordered non-dairy producers to rename their products without reference to "milk" because plant milks do not contain milk from an animal and are therefore misleading to consumers. Plant-based products labelled as cheese or butter were similarly deemed to be misrepresentative (Ledwith, 2017). The U.S. Food and Drug Administration takes a similar stance that products that do not come from a lactating mammal should not be referred to as milk—although this stance has not yet been legally enforced (Atkin, 2018). The market for lactose-free dairy products has grown steadily since the early 2000s (Jelen & Tossavainen, 2003). Early lactose-reducing technologies broke down the lactose molecule using two different enzyme variants, which, in turn, rendered the milk quite sweet because of the high quantities of glucose and galactose. Modifying these early techniques, current lactose-free technologies remove lactose by adding the food enzyme lactase. This addition renders the end product comparable in taste and nutrient composition to conventional (i.e., lactose-containing) 2% milk. As Jelen and Tossavainen (2003) detail, the composition of lactose-free milk is nearly identical to that of partly skimmed milk, with the exception of its lower carbohydrate content (2.8% compared to 4.8%), and, at a 0.01% lactose content, it is legally permitted to be labelled "lactose-free".

In Canada, dairy products with the added enzyme lactase must be labelled "lactose-free" to distinguish them from lactose-containing milk (Canada Food Inspection Agency, 2018). The marketing of lactose-free dairy extends the familiar language of naturalness and healthfulness associated with conventional products. Natrel, a Canadian dairy cooperative in Quebec and subsidiary of larger agricultural parent, Agropur, highlights on their website that

> [t]his is not just lactose free. This is not just rediscovering the great taste of fresh dairy. This is a touch more. This is all of your old favourites, from shortcake to creamy carrot soup. And all of your new favourites too. This is lactose freedom. Taste the possibilities without the discomfort.
>
> (Natrel, n.d.)

Natrel's consumers can choose between skim (i.e., 0%), 1%, 2%, and homogenized (i.e., 3.25%) milks, coffee cream, whipping cream, butter, medium cheddar, chocolate milk, and five different kinds of ice cream, all lactose-free (Natrel, n.d.). Dairyland, a dairy business operating in British Columbia, touts a similar message on their website: "Lactose Free Milk addresses the needs of lactose-intolerant people who have difficulty digesting the natural sugar found in the milk (lactose) but need the essential vitamins and minerals milk provides" (Dairyland, 2017). With yet another technological intervention into dairy, lactose-free milk claims uphold many of the same naturalized truths about dairy's healthfulness. In Natrel's short commercial mockumentary titled *Intolerance: A Lactose Story*, the health benefits of height and strength are centralized alongside "the unique creamy taste that only dairy can provide" (Natrel Milk, 2017).

Whether the production of and reliance on liquid milk, widely used practices of pasteurization, or other alterations to the chemical structure of cow's milk such as homogenization, acidophilus-enhanced, or lactose-free production, the history of milk drinking among humans is a history of what Atkins (2010) refers to as the governmental, scientific, and technological "intrusion into material limits" (p. 160). DuPuis (2002) similarly states that "the story of nature's perfect food is also an industrial story of perfection" (p. 7). Far from natural or universal, the selective framing of milk's history in the human diet speaks as much to the scientific discourses concerning food as it does to the power of industry and advertising in shaping our very assumptions about the materialization of "nature". Akin to other food items of the early industrial era such as wheat, milk "represent[ed] a Victorian and Edwardian anxiety about the uncertainty of nature's intelligibility set against the contemporary urge to believe and capability of science and technology to solve this problem" (Atkins, 2010, p. 192). Illustrated in the marketing of lactose-free dairy campaigns, such as Natrel's mockumentary, scientific interventions into dairy are framed as the solution to lactose-intolerance, the very problem caused by the food itself. In other words, the solution to dairy intolerance is a slightly altered version of the same food item. Naturalized in the commercial history of milk are not only false claims about its inherent pureness as a food source, but also the capacity of science to determine that pureness and, by association, its healthfulness. Despite the deep-seated discourses of naturalization, milk is an industrial food, produced, promoted, and technologically altered to meet shifting health and illness terrains.

Fluid nutricentrisms

The dairy industry has long centralized the language of nutrients in much of its commercial marketing. As Brady et al. (2016) assert, "Milk is not known in static and concrete ways, but is continually constituted and reconstituted by scientific, promotional, and institutional discourse that are always contested and in flux" (p. 92). Intertwined with discourses of naturalization are associations of dairy as a natural source of calcium; protein; magnesium; phosphorous; vitamins A, D, and B12; folate; thiamine; riboflavin; niacin; zinc; and potassium (see, for instance, Dairy Farmers of Canada, 2012). Despite many of these being added after pasteurization and absorption rates being affected by the fat quotient in milk, the value of dairy to the human diet is often measured, discussed, disseminated, marketed, and consumed through nutricentric claims.

This nutricentric framing has not only been produced through industry and marketing efforts, but also reproduced through lay and professional understandings of dairy's health benefits. Wilken and Knudson (2008) highlight that the general public variously draws on well-worn marketing claims. As detailed by their participants, milk is "good for you 'cause it's got calcium, protein and vitamins"; it is "good for your bones and has a lot of calcium"; and it "increases growth...strengthens the nails...and...it makes the bone structure stronger" (Wilken & Knudson, 2008, p. 35). Similar associations were echoed by the dietitians in my study. Faye upholds the value of milk because "there's lots of essential nutrients, [it is] a rich source of calcium, [and contains] easy-to-digest protein". Liz similarly states that "milk products have a lot of good things [...] water and fat and protein and sugar and a bunch of nutrients". Lindsay concurs that "dairy is a great source of calories and protein". And, correspondingly, Spencer dubs dairy "a fantastic source of calcium and a number of other vitamins". Scientized frameworks like these not only reflect, but also actively construct, a nutricentric view of dairy common to industry, professional, and lay understandings.

Noteworthy to my analysis is the degree to which certain nutrients are highlighted over others in relation to changing illness contexts. I use the term "fluid nutricentrisms" to question assumptions that nutricentric truths about the value of dairy to human health are somehow removed from the shifting social and political contexts in which they are produced. As my analysis highlights, nutrient-based health claims associated with dairy are not merely reflective of static biochemical truths. The literal fluidity of milk, I argue, is analogized with the fluidity of its shifting health claims. From the early industrial focus on milk's ability to grow and strengthen to more re-cent claims of aiding weight loss, the nutricentrisms of dairy's health claims continue to morph.

Strength and growth

As DuPuis (2002) documents, by the early 20th century, dairy's authority, which was once linked to the pastoral image of the milkmaid, had shifted to the industrial image of the scientific, male expert. In her words, "this newer overseer of purity, the inspector and the veterinary doctor, replaced the nurturing milkmaid in her generative partnership with nature with the expert male [and] the scientific inspection of [...] the cow" (DuPuis, 2002, p. x). While the milkmaid may have been the epitome of the maternal, the countryside, and purity, early nutricentric framings of dairy focused on dairy's ability to produce growth—the growth of healthy infants into pro-ductive workers, the growth of cities, the growth of industrialization, and the growth of nations.

In the language of scientific racism common to Europeans of the epoch, size and strength were imbued to populations who consumed large amounts of dairy. As Ulysses Hedrick (1933), an American botanist, explains,

> A casual look at the races of people seem to show that those using much milk are the strongest physically and mentally, and the most enduring of the peoples of the world. Of all races, the Aryans seem to have been the heaviest milk drinkers and the greatest users of butter and cheese, the fact that may in part account for the quick and high development of this division of human beings.
>
> (quoted in DuPuis, 2002, pp. 117–118)

It was through this kind of racist rhetoric that milk and dairy became linked to growth and strength—discourses that equally and simultaneously fuelled industrialization and colonization. Building on the shift towards masculine scientific authority and discourses of strength and growth imbued in dairy's "perfect" chemical makeup, some of the earliest print advertisements for dairy, as Maria Veri (2016) observes, relied on images of white male athleticism to drive home the effect of fitness, virility, and strength. Drawing on Victorian discourses of muscular Christianity,[3] masculine athleticism "fused seamlessly with the privilege discourse surrounding the perfection of milk" (Veri, 2016, p. 295). In a burgeoning capitalist system, strong (white) men equalled strong workers, and strong workers equalled strong (colonial) nations.

Vitamin D and rickets

Discourses of strength and growth soon fused with emerging discourses of health and immunity, specifically through the early 20th-century discovery of vitamins. Aware of pasteurized milk's nutritional shortcomings, the dairy industry began fortifying its milk with vitamin D beginning in the 1920s as a way to make it more relevant in the fight against widespread rickets (Veri, 2016). From that point forward, milk touted itself as an important source of vitamin D, despite the fact that it does not occur naturally in the product. Alissa Hamilton (2015) also points out that vitamin D is a fat-soluble vitamin, rendering it virtually useless in fat-free milks. Coalescing with the "vitamania" (Apple, 1996) of the early 20th century, milk secured its place as a protective food item in the face of unseen food-based deficiencies, a claim with which dairy products are most often associated.

Calcium and strong bones

The now ubiquitous links between dairy and calcium appeared in advertising as early as the 1930s. Marion Nestle (2002) points out that at a time when strength in the economy could not be guaranteed (i.e., because of the effects of the economic recession), three glasses a day of milk would maintain strength in one's body. The associations between strengthening the body through calcium consumption and strengthening the nation state intensified in post–World War II North America. As Veri (2016) observes, these associations became one of the main justifications for dairy's central placement in government-issued dietary guidelines. Many have questioned why dairy is its own food group given that all the vitamins and nutrients in dairy (including calcium) can be found in non-dairy food items and given that the majority of the world's population cannot digest the lactose sugars found in dairy (Bhatnagar & Aggarwal, 2007). The seemingly unique role of calcium in dairy was established in the early 20th century and has persisted in much of 20th- and 21st-century advertising.

The 1980s slogan, "Milk Does a Body Good" was a generic statement about the healthfulness of milk, but the message became more specific with the "Got Milk?" campaign starting in 1993, which highlighted milk's ability to promote growth, height, and strong bones and to prevent osteoporosis all because of the presence of calcium (Wiley, 2007). That calcium is also found in non-dairy sources such as fish, tofu, dark green leafy vegetables, and fortified non-dairy drinks has been purposefully obscured by the efforts of dairy lobbying. As Goldberg, Folta, and Must (2002) sum up, milk advertisements have led us to believe that drinking milk is the only and best way to get sufficient calcium. The links between dairy and calcium are so naturalized that it becomes difficult to detangle them and to imagine calcium sources beyond dairy (Gaard, 2013). Over the past 100 or so years of dairy advertising, calcium has been used to promote height, strong bones, dental health, and the prevention of osteoporosis. While dairy was initially promoted to help the growth of infants and children, the focus on calcium also enables it to be marketed to adults and seniors. Despite a decreased ability to digest dairy that many people face as they age, adults and seniors are not outside the nutricentric grip of dairy's health claims, especially with an aging baby boom population in North America.

According to a 2014 study in the *British Medical Journal*, drinking three or more glasses of milk a day may actually increase the risk of bone fractures, especially in women (Michaëlsson et al., 2014). Calling into question the longstanding assumption that a high and consistent intake of dairy is beneficial to bone health, the study speculates that the higher risk of fractures may be due to milk sugars (i.e., lactose and galactose), the very sugars people who cannot digest lactose react to. Another recent study shows that osteoporotic bone fractures are highest in Western countries where people consume the most dairy and calcium and lowest in countries where calcium intake is reduced (Lanou, 2009). While dairy is high in calcium, many claim it is better absorbed through non-dairy food sources such as beans and greens. As A. J. Lanou (2009) details, beans and most greens have an absorption rate of 40% to 64%, compared to milk's 32%, and contain lower sugars and higher fibres, which could explain the higher rates of bone fractures in milk-drinking cultures. The World Health Organization (WHO) explains the calcium paradox by noting, "The adverse effect of protein, in particular animal protein, might outweigh the positive effect of calcium intake on calcium balance" (quoted in Lanou, 2009, 1640s). Instead, the WHO contends that bones are better served by regular exercise, limiting animal protein, and increasing fruit and vegetable intakes.

Protein and weight loss

The most recent shift in dairy's nutricentric framings has been a focus on protein, as it purportedly aids weight loss. Capitalizing on the low-carb manifestos of the late 20th and early 21st centuries that I discuss in Chapter 4, recent dairy ads have decentralized and obscured the language of fat and centralized the presence of proteins. The iconic and long-running Got Milk? ads were recently replaced in the US with a new slogan of "Milk Life", which touts milk's "eight grams of protein" per serving and features regular people doing active things (Kavilanz, 2014). Canadian dairy ad campaigns have also shifted to centralize the role of protein in post-performance recovery and weight loss. With new marketing efforts introducing the tagline, "The original recovery drink" (Chung, 2014), the Dairy Farmers of Canada (DFC) are establishing chocolate milk as post-workout recovery fuel high in protein, sugars, and electrolytes. The DFC began framing chocolate milk as a sport-recovery drink in 2008 and since then sales of chocolate milk have increased 28% (Chung, 2014). Similarly, the DFC's 2012 campaign of "Get Enough" milk links dairy consumption with healthy weight (Dairy Farmers of Canada, 2012).

The weight-loss claims of the dairy industry are curious given that dairy historically was prized as a source of fat not protein (Hamilton, 2015). Since the early 2000s, alongside the shift to low-carb weight-loss movements, the dairy industry has also shifted to promote its products for their ability to help weight loss, despite lack of evidence to support these marketing claims. Four recent reviews and meta-analyses on dairy weight-loss claims conclude that neither calcium nor dairy reliably aids in weight loss, with or without caloric restriction (Lanou & Bernard, 2008). Such findings should not be surprising given that most cheese is 60% to 80% calories from fat, low-fat chocolate milk has the same number of grams of sugar as cola (ounce for ounce), and dairy products are fibre-free, quickening their absorption into blood sugars. While milk may have combinations of nutrients that are useful for sport recovery, so do many other food combinations. What was once a product of growth is now being rebranded as a product of weight loss, despite the nutrient components remaining more or less static. The dairy industry is actively rebranding its image by centralizing protein to fit into the weight-conscious milieu of contemporary culture and by rearticulating new ways in which their product does a body good.

Cancer prevention

Equally audacious, in my opinion, are recent advertising claims that link dairy consumption with cancer prevention. One of the most feared of modern diseases, cancer is the leading cause of death in Canada (Statistics Canada, 2019), the second leading cause of death in the United States (Centers for Disease Control and Prevention, 2017), and the sixth leading cause of death worldwide (World Health Organization, 2018). As part of the 2012 "Get Enough" ad campaign, the Canadian dairy industry circulated images on bus billboards and milk cartons that categorically and simplistically equated dairy consumption with cancer prevention. Against a bright blue background and next to a giant glass of milk still in mid-pour, the white text read, "Help prevent colorectal cancer." On the DFC website, the straightforward framing that dairy consumption "prevents colon cancer" is more equivocated. Allowances that "milk products *could* lower the *risk* of colon cancer" and that "there is enough data to conclude that milk *probably* helps prevent colorectal cancer" (Dairy Farmers of Canada 2014, quoted in Overend, 2016, p. 78) are an obvious refraction of their commercial claim. More recent links between dairy and cancer prevention (including the breast, bladder, prostate, and colorectal kinds) appear on www.dairynutrition.ca, which dubs itself, "The most comprehensive and up-to-date source of scientific information for the health professional on the role of milk products in nutrition and health"—a website run by the DFC. The studies outlined on the website point to the protective roles of vitamin D, conjugated linoleic acid, and calcium as potential cancer-preventing mechanisms found in dairy.

Independent research not outlined on the DFC's website points to contradictory facts about cancer and dairy. Excess calcium from milk may actually increase one's risk of prostate cancer and milk sugars may be linked to a slightly higher risk of ovarian cancer (Abid, Cross, & Sinha, 2014). Dairy consumption has also been implicated in slightly elevated levels of IGF-1, a protein that can contribute to cancer growths (Rogers, Emmett, Gunnell, Dunger, & Holly, 2006). Malekinejad and Rezabakhsh (2015) indicate that the presence of steroid hormones such as recombinant bovine growth hormone (rBGH) in dairy products should be noted as an important risk factor for various cancers in humans. Colin Campbell, professor emeritus of nutritional biochemistry at Cornell University, and his son, Dr. Thomas Campbell, a family physician, also document the links between the consumption of dairy products and breast, prostate, and bowel cancers. In their bestselling book *The China Study*, which reports on 20 years of epidemiological research conducted in 65 regions in China, Campbell and Campbell (2006) contend that "casein, which makes up 87% of cow's milk protein, promote[s] all stages of the cancer process" (p. 131). As such, they maintain that all animal foods, including dairy, should be avoided for optimum human health. Their study, which formed the basis of the 2011 documentary *Forks Over Knives*, has received praise for helping people all over the world prevent illness (Gupta, 2011) and criticism for shaky science and conflicting data (Hall, 2010).

The contradictory health claims of dairy continue to confuse health professionals and the general public alike. Dairy is positioned as preventing cancer on the one hand and contributing to it on the other. In the early 20th century, dairy consumption was supposed to lead to growth and strength, now, a century later, it is promoted as a weight-loss food. From industrial strengthening efforts to neoliberal weight-loss dictums, dairy's nutricentric truths are a roulette wheel of shifting health claims. Depending on the health claim or illness scare du jour, some of dairy's nutrients are centralized while others are downplayed or ignored. Spanning over 100 years of nutricentric health claims, dairy has variously located itself in the North American dietetic imaginary as a fuel for strength and growth, a cure against rickets, a prevention against bone density loss, an aid in weight loss, and a shield against various cancers. Wiley (2007) adds that dairy has also, at various times, extolled itself as a product that can improve skin, help insomnia, prevent aging, and improve immune function. While some of these claims are obvious advertising blights, an overly nutricentric focus on dairy ignores and draws attention away from the broader social and political contexts in and through which these nutricentric claims come to be normalized and naturalized.

Lactose maldigestion and the "unbearable whiteness" of dairy

Taken for granted in and amidst dairy's many oscillating nutricentric claims is that only a small fraction of the world's population is able to digest "nature's most perfect food". This overlooked facet of dairy calls into question its naturalization as a human food item and raises the critical epidemiological question of *who* these nutricentric health claims are purported to help. Andrea Wiley (2004) notes that dairy, because of its white colour, maternal links, and biblical references, carries a positive symbolic association in Western contexts, but that this is far from universal. Greta Gaard (2013) offers a particularly poignant example that when villagers in Columbia and Guatemala were overcome with diarrhea and intestinal cramps after being given powdered milk rations from the United States government, they deduced that the powder must not be for human consumption and used it as a paste to help weather seal their huts. As is well documented across a range of anthropological and medical literatures, only a small fraction of the global population can effectively digest dairy, making it a culturally specific food item that carries with it racialized associations and effects on the constructions of "healthy eating".

Wiley (2004) documents that lactase production (i.e., the production of an enzyme that breaks down milk sugars) for most mammals is highest at birth and during infancy and declines after weaning. Lactase persistence (i.e., the presence of this enzyme into adulthood) is found in populations with long histories of milk and dairy production and the descendants of these populations, namely northern Europeans; South Asians; and herding communities of the Middle East, Arabian Peninsula, and sub-Saharan Africa. However, given that many of these cultures have tended to use *fermented* milk products such as yogurt, *kefir*, and cheese, which are lower in lactose than milk, lactase persistence among these groups is not guaranteed (Wiley, 2004). Lactose intolerance, which is the inability to digest the disaccharide sugar found exclusively in dairy, causing abdominal cramps, diarrhea, and bloating, is therefore the norm for many global populations for whom dairy is not a traditional food item. Roughly one-quarter of Europeans; one-half to three-quarters of Hispanic, Jewish, and Black populations; and close to 100% of East Asian and Indigenous Peoples are considered lactose intolerant (Bhatnagar & Aggarwal, 2007; Wiley, 2004).

Contributing to the West's steadfast reliance on dairy as a natural and healthy food source is its medicalization of lactose intolerance. Sometimes referred to as "lactose maldigesters" or "lactose malabsorption" within Western medical and dietetic discourses, such framings skew and pathologize that which is historically and cross-culturally the norm and treat it as a deviant condition to be solved or fixed (as opposed to just avoiding the food item altogether). For example, the USDA and dairy promotion agencies routinely promote "several easy steps to overcome lactose intolerance" (quoted in Wiley, 2005, p. 510). These "easy steps" include drinking small amounts of milk with meals to slow absorption and working up to larger portions, consuming lower-lactose dairy products like hard cheeses and yogurts, and taking over-the-counter lactase enzymes before consuming dairy products. Similarly, Lactaid Canada (2014), a subsidiary of the pharmaceutical company Johnson and Johnson, admits that "90% of ethnic populations cannot *properly* digest lactose" and have a greater risk of being "lactose *deficient*" (emphasis added to highlight the pathologization of lactose intolerance). Lactaid supplements are advertised to help lactose intolerant people access "the key nutrients found in dairy" and the "several health benefits" associated with dairy consumption (Lactaid Canada, 2014). Given that hypolactasia (i.e., the reduced ability to digest dairy) is the norm for human populations, adults with higher lactase activity would be better described as *hyper*lactasia (Wiley, 2004). Andrea Freedman (2013) similarly comments that it would be more accurate to label people who retain the enzyme that enables dairy digestion as "lactose persistent" than it would be to label those without it as "lactose impersistent" or intolerant. Emphasizing the rarity of lactase production in adults as opposed to the commonality of lactose intolerance may better reflect statistical and cross-cultural rates of dairy tolerance, but it would also undermine dominant discourses of universality and naturalization—core to dairy's construction as a healthful food item and centralized in Western frameworks of healthy eating.

That various cultures are encouraged to whiten their meals with cream sauces and glasses of milk on the side mirrors the whitening of food advice from within dominant Western medical and dietetic paradigms. In her article, *The Unbearable Whiteness of Milk*, Freedman (2013) highlights the ways in which race and whiteness play a part in perceptions of food culture that affect health. She labels the widespread promotion of dairy as a form of food oppression that functions to ignore structural conditions of racial inequality causing health disparities, while simultaneously undermining traditional cultural knowledges of healthy eating, which typically exclude the consumption of dairy, and specifically milk. As outlined in the long history of dairy's production in the West, its centralization to those diets stems from Northern European agricultural practices ported over during colonization and North American economic benefits, alongside industry lobbying and government intervention. Critiquing the colour blindness and universalism of food networks, Julie Guthman (2008) decentres white food advice as dominant and superior. Commonly circulating under the rhetoric of, "If they only knew better", whiteness often establishes itself as the authority on healthy food and contends that poor, marginalized, and racialized populations lack the knowledge to make healthy food decisions. The promotion of dairy to racialized populations only reaffirms this long-standing colonial power dynamic. Highlighting the privileging of whiteness that produces and benefits from particular ways of eating, this can be seen most obviously in Canadian government's promotion of dairy to First Nations, Inuit, and Métis Peoples.

Not only are many Indigenous people lactose intolerant, but as Burnett, Hay, and Chambers (2016) document, the 50-plus-year promotion of Western foods, including dairy, to Indigenous populations further entrenched settler-colonial economic relations, while also impeding and destroying Indigenous food sovereignty. Discussed in Chapter 2, since its inception in 2007, the First Nations, Inuit, and Métis (FNIM) Food Guide includes "milk and alternatives" as one of the four central food items. Specifically, the guide proposes two servings *daily* for children two to three years old; two to four servings for children 4 to 13 years old; three to four servings for teens; two servings for adults (19 to 50 years of age); and three servings for adults over the age of 51 (Health Canada, 2007). The Northwest Territory Food Guide, revised in 2005 and the predecessor to the FNIM Food Guide, also prominently featured dairy as one of the four major food groups, as does the 2011 Nunavut Food Guide (Arceño, 2016; Health Canada, 2007). Food guides such as these formalize and reiterate decades of dairy promotion to Indigenous groups in settler Canada. As Burnett at al. (2016) document, under the Family Allowances Act of 1945—one of the first universal social welfare measures in Canada—Indigenous groups were paternalistically given restrictions on how their money could be spent. Pablum, milk, cheese, and butter were on the list of "acceptable" food—defined, ostensibly, as those high in nutrient value. Equally, in the late 1960s, when the federal government initiated the Food Mail Program as a way to subsidize the cost of transporting nutritious, perishable, commercial foods to northern communities, milk and dairy were again prominent (Burnett et al., 2016). Fresh milk, ultra-high treatment milk, chocolate milk, buttermilk, cream, powdered milk, cheese, processed cheese and spreads, cottage cheese, yogurt, yogurt drinks, ice cream, frozen yogurt, and sherbet were formally endorsed by the Food Mail Program (Grier & Majid, 2010). As Indigenous Peoples were losing rights to their own food sovereignty through the criminalization of hunting and harvesting, they were concomitantly being forcibly fed the health benefits of dairy products by federal government-imposed and subsidized food programs.

Perhaps more than any other food, the ongoing politics of whiteness are evident in the policies and documents that create and promote the healthfulness of dairy to entire populations of people who literally suffer pain and discomfort from its consumption. The brash promotion of dairy is exacerbated by the fact that many First Nations, Inuit, and Métis communities are under "boil water advisories" due to government neglect, therefore minimizing drinking options outside milk, juice, and pop. The colonial use of federal food programs, food guides, and food subsidy programs continue to intersect with growing neoliberal government policies that centralize the needs of domestic agriculture over any genuine interest in population health, which is only heightened for marginalized groups. In 2010 the federal government replaced the Food Mail Program, which was largely a transportation subsidy, with retail subsidies paid directly to northern retail stores. Totalling $54 million per year, these subsidies are given to retail operations and intended to be offered to consumers at the point of purchase; however, the government does little to monitor if subsidies for consumers are actually enforced (Burnett et al., 2016). Running a virtual monopoly of retail stores in the North, the North West Company's profits reached $134.3 million in 2012, two years after the retail subsidy program was initiated; 80% of this capital was generated from the sale of food, including dairy (Burnett et al., 2016).

Despite its centrality to Western food framings, the definitive role of dairy in human health is not always clear or straightforward because health cannot be understood simply as the effect of nutrients consumed. As researchers, professionals, and laypeople alike continue to seek singular nutricentric truths—epitomized in the newly launched Milk Composition Database (Foroutan et al. 2019), which details the more than 2,000 compounds found in cow's milk—this type of nutricentric approach obscures the larger questions we should be asking about dairy in the contemporary moment. My focus on the fluid nutricentrisms of dairy aims to shift the discussion of dairy from one reductively concerned with nutrients to one critically aware of social, historical, and political contexts. In mapping the shifting truths of dairy, I move attention to the subtle and innocuous—yet consistent—ways that the structural epidemics of cancer or heart disease or diabetes or any other contemporary health issue get framed as an individualistic problem of eating rather than a systemic problem of inequality, sustained by government regulations, advertising standards, and/or industry guidelines. If we are continually preoccupied with the nutricentric concern of what to eat or not eat, we fail to look at the structural questions of how, when, why, and under what social and political contexts these health claims are being made and upheld.

An understanding of dairy's shifting health claims highlights the degree to which some of dairy's nutrients are centralized, while others are obscured in relation to shifting health terrains. While the biochemical composition of a 250 ml glass of milk may be relatively fixed (i.e., until new technologies are innovated or new compounds discovered), the nutricentric meanings attached to them are anything but. As Brady et al. (2016) also uphold, "Food is not a static substance which is acted upon by [...] social and historical forces. Rather, the meaning and very materiality of food is brought into being *through* these social forces" (p. 76, emphasis added). The discursive claims of dairy as a "universal", "natural", "perfect", and "healthy" food item are examples of how the social meanings attached to dairy dictate its perceived health benefits. Like most nutricentric food framings in the contemporary era, dairy's health claims establish and uphold a biopolitics of individual responsibility, where one is encouraged to fight a range of structural illnesses one serving of dairy at a time, without any attention to the broader social and political conditions in and through which these claims are produced and in and through which illnesses disproportionately affect some populations more than others. In moving beyond the "got" versus "not" dairy debate, we can focus less on the nutricentric claims therein and more on the cultural contexts that regulate its shifting dietetic truths.

Notes

1 The iconic US dairy industry's "Got Milk" campaign ran between 1993 and 2014 (Tobias, 2014). Reaching universal awareness by the early 2000s, its aim was to link dairy consumption with health, strong bones, disease prevention, and overall vitality through the use of high-priced spokespersons and corporate sponsorship (Manning & Lane Keller, 2004). In addition to professional athletes, models, Hollywood darlings, and fictional characters, in 1998 the ad campaign also featured the then US Secretary of Health and Human Services Donna Shalala, unashamedly putting government support behind commercial enterprise (Nestle, 2013).

2 The phrase "not milk" is attributed to Robert Cohen, an outspoken critic of the US dairy industry. In his book *Milk: The deadly poison*, Cohen (1997) outlines the health risks of milk. He also runs and maintains the website www.notmilk. com, which is a database of anti-milk studies.

3 According to Meyer, Wynveen, and Gallucci (2017), muscular Christianity combined Victorian notions of physical activity with Christian theology. In doing so, athleticism was considered both socially and morally desirable. This also meant that a lack of athletic ability reflected physical and spiritual weakness.

References

Abid, Z., Cross, A. J., & Sinha, R. (2014). Meat, dairy and cancer. *The American Journal of Clinical Nutrition, 100*, 386S–393S. doi:10.3945/ajcn.113.071597

Apple, R. (1996). *Vitamania: Vitamins in American culture*. Darby, PA: Diane Publishing Company.

Arceño, M. A. (2016). *On consuming and constructing material and symbolic culture: An anthropology of pictorial representations of food-based dietary guidelines (FBDGs)* (Unpublished MA thesis). Columbus, OH: Ohio State University.

Atkin, E. (2018, July 20). The war on soy milk. *The New Republic*. Retrieved from https://newrepublic.com/article/150006/war-soy-milk

Atkins, P. (2010). *Liquid materialities: A history of milk, science and the law*. Farnham, UK: Ashgate Publishing.

Bhatnagar, S., & Aggarwal, R. (2007). Lactose intolerance. *British Medical Journal, 30*(334), 1331–1332. doi:10.1136/bmj.39252.524375.80

Bordoni, A., Danesi, F., Dardevet, D., Dupont, D., Fernandez. A. S., Gille, D., ... Vergères, G. (2017). Dairy products and inflammation: A review of the clinical evidence. *Critical Reviews in Food Science and Nutrition, 57*(12), 2497–2525. doi:10.1080/10408398.2014.967385

Brady, J., Millious, V., & Ventresca, M. (2016). Problematizing milk: Considering production beyond the food system. In C. R. Anderson, J. Brady, & C. Z. Levkoe (Eds.), *Conversations in Food Studies* (pp. 75–96). Winnipeg, MB: University of Manitoba Press.

Burnett, K., Hay, T., & Chambers, L. (2016). Settler colonialism, Indigenous peoples and food: Federal Indian policy and nutrition programs in the Canadian north since 1945. *Journal of Colonialism and Colonial History, 17*(2), doi:10.1353/cch.2016.0030

Campbell, T. C., & Campbell, T. M. (2006). *The China study: The most comprehensive study of nutrition ever conducted and the startling implications diet, weight loss, and long-term health.* Dallas, TX: BenBella Books.

Canada Food Inspection Agency. (2018, May 15). Labelling requirements for dairy products. Retrieved from http://www.inspection.gc.ca/food/labelling/food-labelling-for-industry/dairy-products/eng/1393082289862/1393082368941?chap=3

Centers for Disease Control and Prevention. (2017). Leading causes of death. Retrieved July 9, 2019 from https://www.cdc.gov/nchs/fastats/leading-causes-of-death.htm

Chung, M. (2014, April 30). Dairy farmers puts on its game face: The new campaign for chocolate milk calls out the beverage's credentials as a natural, post-recovery workout drink. *Strategy.* Retrieved from http://strategyonline.ca/2014/04/30/dairy-farmers-puts-on-its-game-face/

Cohen, R. (1997). *Milk: The deadly poison.* Englewood Cliffs, NJ: Argus Publishing.

Dairy of Farmers of Canada. (2012, March 12). GetEnough.ca: Why milk products? [Video file]. Retrieved from https://www.youtube.com/watch?v=kb7OTr8h0fU

Dairyland. (2017). Lactose free milk. Retrieved from http://www.dairyland.ca/en/products/milk/lactose-free-milk

DuPuis, E. M. (2002). *Nature's perfect food: How milk became America's drink.* New York, NY: New York University Press.

Fomon, S. (2001). Infant feeding in the 20th century: Formula and beikost. *Journal of Nutrition, 131*(2), 409S–420S. Retrieved from https://academic.oup.com/jn

Foroutan, A., Chi Guo, A., Vazquez-Fresno, R., Lipfert, M., Zhang, L., Zheng, J.,... Wishart, D. S. (2019). The chemical composition of cow's milk. *Journal of Agricultural and Food Chemistry, 67*(17), 4897–4914. doi:10.1021/acs.jafc.9b00204

Freeman, A. (2013). The unbearable whiteness of milk: Food oppression and the USDA. *UC Irvine Law Review, 3,* 1251–1279.

Fulkerson, L. (Director). (2011). *Forks over knives* [Motion picture]. Santa Monica, CA: Monica Beach Media.

Gaard, G. (2013). Towards a feminist postcolonial milk studies. *American Quarterly, 65*(3), 595–618.

Goldberg, J. P., Folta, S. C, & Must, A. (2002). Milk: Can a "good" food be so bad? *Pediatrics, 110*(4), 826–832. doi:10.1542/peds.110.4.826

Grier, S., & Majid, K. (2010). The food mail program: "When pigs fly"—dispatching access and affordability to healthy food. *Social Marketing Quarterly, 16*(3), 77–95.

Gupta, S. (2011, August 25). Becoming heart attack proof [Blog post]. Retrieved from http://thechart.blogs.cnn.com/2011/08/25/becoming-heart-attack-proof/

Guthman, J. (2008). "If they only knew": Color blindness and universalism in California alternative food institutions. *The Professional Geographer, 60*(3), 387–397.

Hall, H. (2010). *The China study* revisited: New analysis of raw data doesn't support vegetarian ideology. Retrieved from https://sciencebasedmedicine.org/the-china-study-revisited/

Hamilton, A. (2015). *Got milked? The great dairy deception and why you'll thrive without milk.* Toronto, ON: Harper Collins.

Harris, M. (1985). *Good to eat: Riddles of food and culture.* Prospect Heights, IL: Waveland Press.

Health Canada. (2007). *Eating well with Canada's Food Guide: First Nations, Inuit and Métis.* Retrieved from http://www.hc-sc.gc.ca/fn-an/alt_formats/fnihb-dgspni/pdf/pubs/fnim-pnim/2007_fnim-pnim_food-guide-aliment-eng.pdf

Health Canada. (2013, September 24). Raw or unpasteurized milk. Retrieved from https://www.canada.ca/en/health-canada/services/milk-infant-formula/raw-or-unpasteurized-milk.html?wbdisable=truehttps://www.canada.ca/en/health-canada/services/milk-infant-formula/raw-or-unpasteurized-milk.html?wbdisable=true

Healthline. (2017a). Foods to avoid with arthritis. Retrieved from https://www.healthline.com/health/foods-to-avoid-with-arthritis#alcohol-and-tobacco

Healthline. (2017b). 12 foods to avoid with IBS. Retrieved from https://www.healthline.com/health/digestive-health/foods-to-avoid-with-ibs#what-to-eat-instead

Jelen, P., & Tossavainen, O. (2003). Low lactose and lactose-free milk and dairy products: Prospects, technologies and applications. *Australian Journal of Dairy Technology, 58*(2), 161–165. Retrieved from https://www.researchgate.net/journal/00049433_Australian_Journal_of_Dairy_Technology

Kavilanz, P. (2014, February 24). Iconic "got milk?" Tagline replaced. *CNN Business.* Retrieved from https://money.cnn.com/2014/02/24/news/companies/got-milk-ad-dropped/index.html

Lactaid Canada. (2014). Are you more likely to be lactose intolerant? Retrieved from https://www.lactaid.ca/whos-affected.html

Lanou, A. J. (2009). Should dairy be recommended as part of a health vegetarian diet? *The American Journal of Clinical Nutrition, 89*(5), 1638S–1642S. doi:10.3945/ajcn.2009.26736P

Lanou, A. J., & Bernard, N. (2008). The dairy and weight-loss hypothesis: An evaluation of the clinical trials. *Nutrition Review, 66*(5), 272–279. doi:10.1111/j.1753-4887.2008.00032.x

Ledwith, M. (2017, June 14). Soy milk? You can't call it milk, says the EU: Order to rename dairy substitutes because they do not contain produce from an animal. *Daily Mail.* Retrieved from https://www.dailymail.co.uk/news/article-4605566/Soy-milk-t-call-milk-says-EU-court-justice.html

Malekinejad, H., & Rezabakhsh, A. (2015). Hormones in dairy food and their impact on public health. *Iranian Journal of Public Health, 44*(6), 742–758.

Manning, J., & Lane Keller, K. (2004, January/February). Got advertising that works? How the "got milk?" campaign shook consumers out of their milk malaise. *Marketing Management*, 16–20. Retrieved from https://studylib.net/doc/8870338/got-advertising-that-works%3F

Meyer, A., Wynveen, C., & Gallucci, A. (2017). The contemporary muscular Christian instrument: A scale developed for contemporary sport. *International Review for the Sociology of Sport, 52*(5), 631–647. doi:10.1177/1012690215611392

Michaëlsson, K., Wolk, A., Langenskiöld, S., Basu, S., Lemming, E. Melhus, H., & Byberg, L. (2014). Milk intake and risk of mortality and fractures in women and men: Cohort studies. *British Medical Journal, 349*(7981), 6015–6029. doi:https://doi.org/10.1136/bmj.g6015

Myers, A. (2013, April 12). The dangers of dairy. Retrieved from https://www. amymyersmd.com/2013/04/the-dangers-of-dairy/

Natrel. (n.d.). Lactose free. Retrieved from https://www.natrel.ca/en/products/ lactose-free

Natrel Milk. (2017, October 10). *Intolerance: A lactose story* [Video file]. Retrieved from YouTube https://www.youtube.com/watch?time_continue=2&v=odVd1IOi7cI

Neslte, M. (2013). *Food politics: How the food industry influences nutrition and health.* Berkley, CA: University of California Press.

Osborn, D. (2017). Therapeutic management of the phlegmatic humor and temperament. Retrieved from http://www.greekmedicine.net/Principles_of_Treatment/ Managing_the_Phlegmatic_Temperament.html

Overend, A. (2016). Mothering discourse and the marketing of dairy as a cancer-fighting food. In F. Pasche-Guignard & T. Cassidy (Eds.), *Mothers and food: Negotiating foodways from maternal perspectives* (pp. 73–87). Bradford, ON: Demeter Press.

Rogers, I., Emmett, P., Gunnell, D., Dunger, D., & Holly, J. (2006). Milk as a food for growth? The insulin-like growth factors link. *Public Health Nutrition, 9*(3), 359–368. doi:10.1016/j.nupar.2015.09.006

Schmid, R. (2009). *The untold story of milk, revised and updated: The history, politics and science of nature's perfect food: Raw milk from pasture-fed cows* (2nd ed.). White Plains, MD: New Trends Publishing.

Sethi, S., Tyagi, S. K., & Anurag, R. K. (2016). Plant-based milk alternatives an emerging segment of functional beverages: A review. *Journal of Food Science and Technology, 53*(9), 3408–3423. doi:10.1007/s13197-016-2328-3

Statistics Canada. (2019). Leading causes of death, total population, by age group. Retrieved July 9, 2019 from https://www150.statcan.gc.ca/t1/tbl1/en/ tv.action?pid=1310039401

Tan, J. (2017, November 6). Foods to avoid when you have a cold or flu. Retrieved from https://www.avogel.co.uk/health/immune-system/foods-to-avoid-if-you-have-a-cold-or-flu/

Tobias, C. (2014, March 21). R.I.P got milk? Marketing campaign ends after 20 years and plummeting milk sales. Retrieved from https://www.canadianbusiness.com/ companies-and-industries/r-i-p-got-milk-marketing-campaign-ends-after-20-years-and-plummeting-milk-sales/

Tompkins, K. W. (2009). Sylvester Graham's imperial dietetics. *Gastronomica: The Journal of Food and Culture, 9*(1), 50–60. doi:10.1525/gfc.2009.9.1.50

Tompkins, K. W. (2012). *Racial indigestion: Eating bodies in the 19th century.* New York, NY: New York University Press.

U.S. Food and Drug Administration. (2012, August). *The dangers of raw milk.* Retrieved from https://www.fda.gov/downloads/Food/FoodborneIllnessContaminants/UCM239493.pdf

Valenze, D. (2011). *Milk: A local and global history.* New Haven, CT: Yale University Press.

Veri, M. (2016). Got athletes?: The use of male athlete celebrity endorsers in early twentieth-century dairy-industry promotions. *Journal of Sport History, 43*(3), 290–305. doi:10.5406/jsporthistory.43.3.0290

Wiley, A. (2004). "Drink milk for fitness": The cultural politics of human biological variation and milk consumption in the United States. *American Anthropologist, 106*(3), 506–517. doi:10.1525/aa.2004.106.3.506

Wiley, A. (2007). Transforming milk in a global economy. *American Anthropologist, 109*(4), 666–677. doi:10.1525/AA.2007.109.4.666

Wiley, A. (2011). *Re-imagining milk: Cultural and biological perspectives* (1st ed.). New York, NY: Routledge.

Wiley, A. (2014). *Cultures of milk: The biology and meaning of dairy products in the United States and India.* Cambridge, MA: Harvard University Press.

Wilken, L., & Knudson, A-C. L. (2008). Milk, myth and magic: The social construction of identities, banalities, and trivialities in everyday Europe. *Kontur— Tidsskrift for Kulturstudier, 17,* 33–40.

World Health Organization. (2018). Top 10 causes of death. Retrieved from https://www.who.int/news-room/fact-sheets/detail/the-top-10-causes-of-death

4 Wheat
Global staple, modern-health scourge

Of the many dietary debates circulating in contemporary culture, perhaps none seems more pervasive or contradictory than those about wheat.[1] A longstanding global staple for most of the world's population, on one hand, and a perceived dietary threat for a growing number, on the other, the healthfulness of wheat and the many food items that contain it continues to be debated by scientists, scholars, journalists, and popular authors, as well as the general public. Gluten-free products have sprouted in grocery stores and on menus at the same time that wheat has proliferated as an additive in previously wheat-free food items, including processed meats, sauces, and snacks. Despite current concerns about gluten circulating in television talk-show deliberations and celebrity fad diets, in his incisive history of white bread, Aaron Bobrow-Strain (2012) notes that "for as long as western culture has existed, it has been accompanied by anxieties about wheat and its refining" (p. 78). As far back as the classical period, Plato questioned the impact of refined wheat on society's moral health. The health reformers of the late 19th century similarly vilified refined flour and industrialized milling. As this chapter will trace, historic debates concerning wheat continue to inform contemporary arguments of the benefits and risks of going grain-free and contribute to the symbolic role wheat plays in neoliberal Western dietetic culture.

While wheat has remained a staple ingredient in deliberations about health for a very long time, the specific terms in and through which it has been vilified (and glorified) have shifted. I sketch the genealogy of wheat's social and biopolitical history in North American dietetic history to understand the sticky terms of current gluten-free claims. In line with my claims on dairy in Chapter 3 and meat in Chapter 5, this chapter will not provide any definitive answer to the contemporary question of wheat's healthfulness but will instead question the health debates that surround it. As some of my interviewees also maintain, there is no singular dietetic truth to wheat's healthfulness. As Emma puts it, "Wheat can be part of a healthy diet as long as people don't have any issues digesting it"; for Fay, "it would depend on the context [of the diet]"; likewise for Jade, "it would depend on the quality of the soil and the flour"; and in Liz's words, "I think it depends. I don't think we understand gluten intolerance very well". Moving away from singular food truths and towards an understanding of the ways that contemporary wheat-free debates come to be situated offers insights about the current workings of food- and eating-based politics. To explore what these debates convey about social fears and regulations, I first summarize the pro- and anti-wheat debates. Next, I situate these debates in relation to wheat's racialized, classed, and gendered histories. And finally, I assess the discursive workings of current anti-wheat arguments as they fuel contemporary articulations of fatphobia.

Global staple

One of the recurrent themes in my conversations with dietitians about wheat is how central it is to many global diets. Jessica points out that there is some sort of starch or grain basis globally. Bianca holds up the Mediterranean diet, which has been recently touted for its health benefits, as an example of a traditional diet with wheat as a dietary staple. The etymologies of the word "wheat" further confirm its longstanding role in a range of human diets.

The English word "wheat" stems from old Germanic "*hwaitja*" or "*hwita*", meaning "white". Referring to the lightness of the grain and its associated ability to produce white bread, the linguistic connection to whiteness is also evident in the current German "*weizen*" and the Danish "*hvede*" (Bjørnstad, 2012). The colour of the grain, however, is not emphasized in all cultural etymologies of the word. The Latin word "*frumentum*" and the French "*blé*", simply mean the most common of the grains. The Russian "*psenica*", the Czech "*psenice*", the Slovakian "*psenica*", and the Polish "*pszenica*" derive from the word "millet", meaning "to pound" or "to crush", referring to the labour-intensive practice of extracting the edible part of the plant (Bjørnstad, 2016). In Middle Eastern contexts, the Akkadian "*kibtu*" means heavy grain, and in Egyptian, the word for bread is "*eish*", referring broadly to life and living (Bjørnstad, 2012). The significance of bread to Western and global culinary histories is further evident in common expressions, phrases, and religious practices. To "break bread" is to share a meal, "bread and butter" refers to something quotidian, and Judaism, Christianity, and Islam all use bread-based traditions in food and religious ceremonies.

Wheat was domesticated roughly 10,000 years ago during the transition from hunter-gatherer to agricultural societies (Shewry, 2009). For much of this period, wheat had been a staple food for Europe, the Middle East, West Asia, North Africa, and eventually North America. Today, wheat occupies more land area than any other commercial crop and continues to be a vital grain and protein source for humans (Bjørnstad, 2016; Curtis, 2002). Counted among the big three of cereal crops, alongside rice and maize, wheat is considered unique. It is a high-yielding crop, readily harvested, and effectively stored for long periods. It is also durable and grows in a range of temperate conditions, in northern to southern regions of the globe, and is a grain that yields significant genetic diversity (Shewry, 2009). Moreover, because of the elasticity of glutemins and the viscosity of gliadins, what is referred to as the "viscoelasticity" of wheat, it is the grain most commonly used for bread and cereals worldwide (Bjørnstad, 2016). The malleability of wheat is evident to bakers who stretch and shape dough into its many forms and in the seemingly endless array of leavened and unleavened breads, pastas, pastries, and cakes found worldwide. In addition to wheat- and bread-based products, wheat has also been widely used as a binder for processed and pre-packaged foods and as a filler for animal feed production (Shewry, 2009).

Introduced to the United States by European colonists as early as the 15th century, wheat did not initially thrive as well as the native corn (Tompkins, 2009). Introduced a few centuries later to Canada, the grain eventually flourished in parts of the US and the southern regions of the Canadian prairie provinces. The dryness of the latter is an important factor in producing a high-quality, protein-rich grain. The grain, as André Magnan (2016) details, has "transformed the Canadian Prairie landscape into fields of gold, tying farming households to distant export markets" (p. 10). The quality and success of Canadian wheat has withstood what Magnan (2016) summarizes as three changes to its export: the UK food regime from 1872 to 1914, the mercantile industrial food regime from 1945 to 1972, and the current corporate food regime from 1995 to the current moment. Canada remains in the top ten wheat producing countries in the world, with the majority being produced on the prairies (World Atlas, 2019). In 2018, wheat was the largest estimated seeded area in Canada, just ahead of canola and far ahead of pulses, soybeans, barley, and oats (Canadian Wheat, 2018). However, Emily Eaton (2013) notes that since the privatization of the Canadian Wheat Board in 2015, the quality of Canadian wheat is threatened by agricultural giants such as Monsanto because of monocropping and mass pesticide practices.

Further factors contributing to wheat's popularity, success, and widespread use include its mild taste, especially when compared to strongertasting grains like rye; its ease of digestibility, in contrast to denser grains like barley; and its quick energy boost because of easily metabolized sugars (Bjørnstad, 2016). In its common forms of bread, cereal, and pasta, wheat is also a portable and convenient food source and supplier of antioxidants, fibre, and selenium (Bjørnstad, 2012). My interviewees corroborated wheat's beneficial fibre and nutrient properties (Heidi, Eliete), which are also echoed by health authority agencies in Canada. The 2007 Canada Food Guide recommended between six and eight servings of grain per day for adults, and the 2019 Food Guide urges Canadians to fill roughly one-quarter of their plate with whole grains. The Heart and Stroke Foundation of Canada (2018) pushes grains as a healthy source of dietary fibre. And the Canadian Cancer Society (2018) touts the benefits of whole grains for the presence of antioxidants (vitamins A, C, E, and selenium) and phytochemicals (carotenoids, flavonoids, indoles, and isoflavones). Despite wheat's status as a staple grain, its versatility and durability as a crop and food item, its central place in Western culinary food traditions, and it being a convenient source of vitamins and nutrients, a growing number of popular authors, opinion leaders, alternative health practitioners, some mainstream medical doctors, and a few outspoken celebrities question the necessity of wheat in human diets.

Modern health scourge

Critics of this age-old grain carry reservations about its transmutation through various industrialized food practices, from milling and refining to genetic modification, and its possible connection to the rise of non-communicable illnesses in the post–WWII era. Such illnesses are sometimes lumped under chronic diseases of civilization and include diabetes, heart disease, cancers, and the "obesity epidemic" (see endnote 3, Chapter 2). Despite the near ubiquitous framing of "obesity" as an epidemic in medical circles and mainstream discussions of health, critical health and fat studies scholars reject "obesity" as an illness because it medicalizes fatness and falsely establishes it as a condition to be fixed or cured (Guthman, 2011; Lebesco, 2004).

Central among the anti-carb voices of the late 20th and early 21st centuries is cardiologist and weight-loss enthusiast Dr. Robert Atkins. After receiving his medical degree from Cornell University in 1955, Atkins began researching weight-loss nutrition in the 1960s. In 1972 he published the first edition of *Dr. Atkins' Diet Revolution*, a commercial weight-loss program focused on carbohydrate restriction (Gilman, 2008). In 1992 he published *Dr. Atkins' New Diet Revolution*, an updated version of the original food plan, and, by 1997, the book became a bestseller and remained so for five years (Gilman, 2008). Promoting a high-fat, high-protein, low-carb diet, the Atkins's nutritional approach gained widespread popularity in the early 2000s. Atkins's claim was that the public had been misinformed by mainstream dietary advice about the risks of saturated fats and the benefits of carbohydrates: "These twin epidemics, obesity and diabetes, were clearly a result of the low-fat high-carbohydrate diet that was being preached to the public as gospel" (Atkins, 1992/2002, p. xii). The over-consumption of bread and other carbohydrates were the root cause of weight gain and illness and the Atkins's empire of cookbooks, meal plans, vitamins, and frozen meals promoted his low-carb philosophy.

Spreading the anti-carb manifesto with "one best-selling book after another", Atkins's hypothesis was that carbohydrates were the problem with our growing waistlines, not fats (Taubes, 2008, p. xix). In a widely debated *New York Times Magazine* article, Gary Taubes (2002) asks, "What If It's All Been a Big Fat Lie?", arguing, in line with Atkins, that carbohydrates are driving up people's insulin levels and causing weight gain. Taubes points specifically to refined carbohydrates (i.e., sugar, flour, and white rice) that have been machine-processed to make them more easily digestible—a process that has also removes many of the important fibres that allow for them to be more slowly metabolized. In 2008 Taubes expands his thesis in *Good Calories, Bad Calories*, which claims that misguided science has led to misconceptions about the relation between carbohydrates and heart disease, high blood pressure, cancer, dementia, diabetes, and "obesity". When it comes to determining the ideal diet, he says, we have to "confront the strong possibility that much of what we've come to believe is wrong" (Taubes, 2008, p. xx). When North Americans have eaten less saturated fat and yet the incidence of "obesity" and diabetes has increased, he urges us to reconsider the dominant low-fat, high-carb dietary wisdom.

In 2002 health scientist Loren Cordain published *The Paleo Diet* and launched what would become the paleo movement, deepening the low-carb Atkins approach. Based on gastrointestinal research first conducted in the 1970s (Voegtlin, 1975) and developed in the 1980s (Eaton & Konner, 1985), *The Paleo Diet* is premised on what our hunter and gatherer stone age ancestors ate: wild plants and animals. The paleo diet restricts processed foods; refined sugars; most dairy; starchy vegetables like potatoes; legumes such as beans and lentils; and all grains, including wheat, rice, corn, oats, and barley. Non-starchy fruits and vegetables, nuts, seeds, and meats are allowed. Like Atkins, Cordain (2002) attributes the rise of diseases to the Agricultural Revolution of the mid-17th to late 19th centuries and resultant influx of grains in the human diet. The paleo diet, along with its derivative ketogenic diet,[2] has garnered support from scientific, medical, and lay communities alike as an effective weight-loss and anti-diabetes program (McNeil, 2010), while others are skeptical of its long-term health consequences (Ornish, 2015) and its socio-cultural implications (Reeves, 2017).

By the late 2000s, professional athletes, Hollywood darlings, and television talk-show hosts were all praising the accolades of eating gluten-free: the diet is claimed to aid weight loss, unleash reservoirs of energy, stabilize insulin levels, and even promote mental sharpness (Bobrow-Strain, 2012; Taubes, 2008). The growing chorus of anti-wheat spokespersons contended in some way that North American wheat and the quantities in which we eat it are making us needlessly heavy, tired, and sick. Sceptics of wheat and gluten point to "modern plant breeding, [...] pesticides, endocrine disruptors, high speed dough handling, industrial fermentation, genetic modification, or the unbridled use of wheat-based additives in foods that never before contained gluten" (Bobrow-Strain, 2012, p. 77). While it might be easy for some to dismiss these arguments under broad strokes of "dietary quackery", some of the same rationale was invoked by the dieticians I consulted for this research. As Eliete states, "I think the push from the government to eat lower fat foods has resulted in people eating or overemphasizing grains, including wheat, in their diet". Andrea takes a categorical stance against eating North American wheat because it "has been genetically modified; heavily sprayed with glyphosate, which is toxic for our bodies; and the soil [it grows in is] depleted, which doesn't really add any nutritional value to our lives". Victoria similarly raised concern over the chemical spraying of most North

American wheat fields: "I can't image that the glyphosates aren't causing some kind of alteration in our gut flora". And Brianna mentioned the large spectrum of gluten intolerance, the high protein (i.e., gluten) component of North American wheat as compared to most European wheat, and the level of processing affecting inflammatory conditions such as rheumatoid arthritis.

Another prevailing popular voice in the contemporary wheat-free movement is Dr. William Davis, a cardiologist and author of the bestselling book *Wheat Belly*. Citing a host of successful cases, Davis (2011) asserts that by removing wheat from one's diet, a person can reverse illness, including diabetes and its associated health consequences. He blames wheat for causing diabetes because it increases blood sugar more than any other carbohydrate and provides opportunity for glucotoxicity, lipotoxicity, and inflammation, accelerating the rate of kidney dysfunction, dementia, arthrosclerosis, and arthritis (Davis, 2011, pp. 115, 139). At the same time, wheat triggers a cycle of insulin-driven satiety and hunger, matched by euphoria and withdrawal for wheat- and gluten-based foods. He claims that it is unlikely that wheat causes autism, ADHD, and schizophrenia, but does assert that the grain is associated with inflaming their symptoms (Davis, 2011, p. 48). Like other wheat sceptics, Davis laments the ubiquity of wheat in a range of commercially sold food products and the effects of industrialized agricultural practices, including genetic modification. As he asserts, "Modern wheat, despite all the genetic alterations to modify hundreds, if not thousands, of its genetically determined characteristics, made its way to the worldwide human food supply with nary a question surrounding its suitability for human consumption" (Davis, 2011, p. 30). Unlike the more robust wheat of generations past, he contends that today's wheat is "a perfect, chronic poison" (Davis, quoted in Cochrane 2013).

David Perlmutter (2013), neurologist and author of *Grain Brain*, echoes the refrains of Davis (2011), Atkins (1981, 1992/2002), Cordain (2002), and some of the dietitians I interviewed for this research. He states that "through modern hybridization and gene-modifying technology, the 133 pounds of wheat that the average American consumes each year shares almost no genetic structural or chemical likeness to what hunter gatherers might have stumbled upon" (Perlmutter, 2013, p. 8). It is the shift in the type of wheat consumed combined with the shift from low-fat to high-carb diet that he argues is the origin of many current health issues linked to the brain, including chronic headaches, insomnia, anxiety, depression, epilepsy, movement disorders, schizophrenia, and ADHD. Refuting the commonsense assumption that wheat is part of a balanced diet, he argues that the human dietary requirements for carbohydrates is virtually zero: we can survive on a minimal amount of carbohydrates, but we can't go long without fat (Perlmutter, 2013). Citing clinical cases, he maintains that the single dietary shift of removing grain from one's diet can counteract the effects of depression, chronic fatigue, type 2 diabetes, obsessive-compulsive behaviour, brain fog, and bipolar disorder (Perlmutter, 2013).

A perusal of news headlines, book reviews, and dietary blogs reveals that the contemporary wheat- and grain-free movements have generated equal amounts enthusiasm and scorn. Public and scientific support for *Dr. Atkins' Diet Revolution, The Paleo Diet, Wheat Belly,* and *Grain Brain* corroborate some of what the authors contend. In Amazon's and Indigo's online review forums, readers added comments such as, "Reads like science fiction but has a ring of truth" (Beatrice, 2011), "You can't deny the truth when you test it out for yourself" (Angel, 2016), and "Now I am looking at bread a lot differently and leaving it where it belongs…in the package" (Shikes, 2012). Others mentioned that even though they still are eating wheat, they have an increased interest in the subject and are more inclined to think that the 2007 Canada Food Guide recommendations for wheat were misguided. Scientific sources such as the *Journal of the American Medical Association* confirm that premenopausal overweight women assigned to the Atkins diet, which had the lowest carbohydrate intake, lost more weight at 12 months than women assigned to other weight-loss diets (Gardner et al., 2007). Likewise, a 2013 study published in the *American Journal of Clinical Nutrition* finds that low-carbohydrate, high-protein diets are an effective means of diabetes management (Olubukola, English, & Pinkney, 2013). Equally, a growing number of Canadian physicians are supporting the paleo and keto diets as a means of weight loss and disease maintenance (Canadian Clinicians for Therapeutic Nutrition, 2019).

Furthermore, a 2014 Canadian study shows that gluten is the fastest growing dietary intolerance category (Agriculture and Agri-Food Canada, 2014). As the results of this study summarize, roughly 1% of the Canadian population consumes gluten-free products because of celiac disease;[3] 6% because of non-celiac gluten sensitivity; and a growing 22% for reasons such as digestive health, nutritional value, weight loss, healthier skin, joint pain relief, mental function, stress relief, cleansing regimen, depression relief, asthma, and allergies (Agriculture and Agri-Food Canada, 2014). Both respiratory and food allergies to wheat are not new to the grain's history. Respiratory allergy, or what was colloquially referred to as "baker's asthma", has been documented since Roman times and remains a major occupational allergy for those handling flour (CBC News, 2018). Wheat is also routinely classified one of "the Big Eight" of food allergens, accounting for 90% of all allergic reactions, and, in extreme cases, anaphylaxis (Shewry, 2009). Celiac disease is the most documented of these allergic reactions, but other autoimmune, digestive, and neurological responses to wheat include dermatitis, eczema, rheumatoid arthritis, irritable bowel syndrome, and autism (Elder, 2008; Fasano, Sapone, Zevallos, & Schuppan, 2015).

Despite a great deal of professional and lay uptake of the arguments of going wheat- and grain-free, Atkins, Cordain, Davis, and Perlmutter have also been critiqued for their exaggerated writing styles, handpicked data, money-making empires replete with cookbooks and pre-prepared meals, and overly simplistic approaches to complex human health problems. Alan Levinovitz (2015a, 2015b) summarizes some of these critiques. In his popular press book *The Gluten Lie*, with an equally exaggerated writing style, he aims to expose Davis and Perlmutter as sensationalists not scientists:

> At best *Wheat Belly* and *Grain Brain* are collections of unfounded speculations, cherry-picked data, and overconfident hypotheses. At worst they're tantamount to medical malpractice, snake oil in the literary form that should earn the authors the censure of their professional peers.
>
> (Levinovitz, 2015a, p. 28)

Some sources go as far as to say that the gluten-free diet is not only misguided quackery but may in fact contribute to weight gain, increased cardiovascular risk, and other long-term health effects (Katz, 2003; Lagiou et al., 2012).

Whether pro- or anti-wheat, critiques on both sides of the debate ultimately get mired in the singular truth-seeking approach to food that I am deeply wary of. Throughout his popular press book, Levinovitz (2015a) uses phrases such as "getting at the truth" (p. 31), "reveal the missing superstition behind fears of gluten fat, sugar, and salt" (p. 22), and "lies about gluten" (p. 32)—this last example is also the title of his book. This language essentializes and upholds the presumption that there are singular, discernable, and universal truths about wheat as a food item. My aim of a post-singular truth approach to food studies is to shift some of the categorical thinking epitomized in the pro- and anti-wheat discussions. In shifting the "good/ bad", "healthy/unhealthy" binaries of contemporary wheat- and grain-free discourse, I open up wheat's fluctuating health-based arguments to highlight the broader social regulatory practices that these discourses invariably invoke. Referring to what Elspeth Probyn (2013) calls the "biocultural entanglements" of food and eating practices, which highlight "the power relations integral to eating" (p. 333), I turn now to the raced, classed, gendered, and size-based discourses of wheat's sticky social history, to better contextual its place in contemporary food biopolitics and social discourses of healthy eating.

Wheat's sticky socio-political discourses

A critical reading of wheat's historical and contemporary cultural contexts highlights the anxieties of the time as well as the social regulation of those anxieties. The aim here is to decentralize the "what" of wheat's role in the human diet by centralizing the social and political discourses attached to wheat-based health framings. Drawing on the work of Kyla Wazana Tompkins (2012), I first trace what she terms the "imperial dietetics" (p. 53) of wheat's early industrial history, a movement that conflated food and racial purity and that is still evident today. Next, I frame the wheat- and grain-free health movements of the late 20th and early 21st centuries as an extension of the fatphobic dietetic discourses of the 1980s. Through this transition, a fear of dietary fat is replaced with a fear of carbs, but the underlying negative association of fatness with disease is upheld and normalized, while the broader systemic and structural causes of illness are ignored. In both cases of racial and thinness purity, the question of pro-wheat versus anti-wheat is much stickier and more complex than the grain's nutricentric components and must therefore be understood and unpacked within wider social and discursive constructs.

Imperial dietetics and racial purity

As Tompkins (2009, 2012) has effectively argued, far more than being just a staple grain, wheat is inextricable from the history of European settlement and, with it, colonial ideology. Recalling that corn, not wheat, was native to North America, Tompkins (2012) explains that wheat's migration as a dietary mainstay alongside European expansion was made possible by what she terms the "imperial dietetics" (p. 53) of the antebellum period. Imperial dietetics refers to the white nationalist ideologies that conflated food purity moments of the mid-to-late 19th century with early racist, classist, and sexist social purity reforms. Producing an influential and enduring conceptualization between diet and moral well-being in 19th-century America was the evangelic minister, social reformer, and outspoken advocate of dietary reform Sylvester Graham (1794–1851).

With a relatively brief stint in the public eye, Graham's rigid dietetic treatise was both pervasive and enduring. Not only was the current-day Graham cracker created with his health principles in mind, but he motivated other 19th- and 20th-century health reformers, including the likes of cereal giant John Harvey Kellogg (Gilman, 2008). Graham was a chronically ill child, later turned vegetarian, who became a campaigner in the Temperance Movement. Responding to the influx of prepared foods with religious fervour, in his *Treatise on Bread and Breadmaking*, Graham (1837) upheld homemade, whole-wheat bread as the ideal food (Tompkins, 2009). According to Graham, eating whole-wheat bread, along with avoiding alcohol, meat, spices, sugar, caffeine, and refined white flour, was a way to temper digestive inflammation and sexual voracity (Thompkins, 2012). Reciting humoural language and logic, Graham regarded farinaceous foods, with wheat foremost among them, as essential to the chaste and optimal functioning of the body because they provide "enough roughage to keep [the body] both cool and fit, preventing overheating and enervation" (Tompkins, 2009, p. 54). In *A Lecture to Young Men on Chastity*, Graham (1838) contended the cooling function of bread was the cure for many social ills— namely, the epidemic of masturbatory young men, but also that of perceived racial impurity threatening the white symbolic order of imperial ideology.

Tompkins (2009) articulates the ways in which Graham's imperial dietetics "uncover slippery relationships among eating, domesticity, race, and national formation" (p. 51). Bread for Graham signified not only proper domestic order—keeping women tethered to the labour of homemade loaves within the domestic sphere[4]—but also the normalization and expansion of white imperialism at the heart of empire. As a colonial food, wheat served as a far more important symbol of the imperial project than corn. Indeed, most of the foods marked as threatening to Graham's project were those deemed "foreign" (i.e., spices, coffee, sugar, tea, and wine), while the remedy for social, racial, and moral degeneracy rested in the consumption of newly domestic food items (i.e., wheat and dairy). In classic imperial logic, the solution to problems that were created through colonization were themselves colonial, ensuring once again the white man's place at the centre of nation building and expert dietary advice. Graham's food reform effectively established a parallel between the individual body, the domestic sphere, and the newly defined nation, where each served to regulate the other through dietetic and colonial rule (Tompkins, 2012). The body became what it ate and the strength of the nation was indexed by the strength of what was eaten. Conflating dietary reform with early nationalist racism, "ingesting more bread, [Graham] promised, would produce healthy bodies and homes and guarantee the United States' place in the pantheon of civilized nations" (Tompkins, 2009, p. 51). Extending Grahamite sentiment on wheat's role in the growth of empire, Dr. Woods Hutchinson, a prominent American health writer, denounced that

> one need only compare strapping, tall Americans with specimens from any rice- or brown- bread-eating nations. In strength, valor, and intelligence the American surpasses them all. [...] White flour, red meat, and blueblood are all the emblems of global conquest.
>
> (quoted in Bobrow-Strain, 2012, p. 96)

The individual body and the body of the colonial nation state were simultaneously sustained by the consumption of wheat.

Despite Graham being a staunch critic of white bread, his food purity politics paradoxically extended into the mass adoption of industrialized, store-bought white bread. With the conflation between food and social purity that Graham articulated, dirty or impure bread came to be symbolically linked with immigrants who had a palate for dark, unrefined flour, as well as with "mom's home-baked loaf" (Tomes, 2000). In order to sell the newly configured industrial loaf, industrial bakers had to cast serious doubt on longstanding domestic bread-making traditions, and they did so in part through Grahamite food racism and in part through the thinly veiled discourses of scientific sexism. As the National Association of Master Bakers articulated in 1914,

> The average housewife of today who bakes bread is living in the dark; she is ignorant of what the up to date method of baking consists; She has to be educated, the same as a child is educated to eat from a plate—the only difference being that our task is far harder than teaching a child, whose mind is receptive to instruction and learning.
>
> (quoted in Bobrow-Strain, 2013, p. 279)

By contrast, the uniformly baked, meticulously sliced, whiter-than-white industrial loaf became a "visual spectacle of purity, hygiene, and progress" (Bobrow-Strain, 2013, p. 280). At a time when bread-borne diseases were disproportionately blamed on immigrant and home bakeries, industrial bread promised cleanliness and hygiene (through xenophobia and misogyny) as its greatest selling points, fostering food purity discourses as part and parcel of sliced bread.

Despite its commercial success, the public acceptance of industrial white bread was far from unanimous, and this remains the case today. Bobrow-Strain (2007) explains that by the 1920s,

> The modest, ordinary loaf of white bread had been accused of some extraordinarily immodest deeds: the leading cause of anemia, cancer, diabetes, criminal delinquency, tuberculosis, polyneuritis, gout, rheumatism, childhood blindness, overstimulated nervous systems, acidosis, morbidity of mind and body, and white race suicide.
>
> (p. 45)

While industrial bakers had almost exclusively moved baking from the domestic to the public sphere, criticisms about the quality of the store-bought bread began to mount (Kaplan, 2006). Industrial bread was, according to the dietary experts of the day, "flavorless, adulterated, and unvital sacks of hot air", "so clean a mealworm can't live on it for want of nourishment", and "white and waxy as the face of a corpse" (quoted in Bobrow-Strain, 2007, p. 47). Those critical of industrial white bread blamed corrupt corporations, mechanical milling, lazy consumers, bad mothers, dirty immigrants, and industrialization itself.

Indexed through wheat's historical imperial dietetics, the adoption, promotion, and regulation of bread as a staple food in North America was not accomplished without the use of wider social discourses concerning gendered racisms and classisms. Colonial pursuits were accomplished, in part, through dictums of wheat as a healthful food choice, not only in place of other grains, but also through imperial, industrial means of production. While Tompkins's (2009, 2012) analysis focuses on the development of US nationalist propaganda, a parallel pattern existed in Canada. European homesteading families were often credited with "building the West" through grain expansion. Åsmund Bjørnstad (2012) points out that the Marquis Wheat 1909 postage stamp, released in Canada in 1988 as part of a series on Canadian innovations in energy, food, research, and medicine, is an example of the colonialist, nationalist wheat ideologies that celebrated and promoted "the building of the West", while erasing colonial practices of Indigenous land and the food displacement embedded in grain expansion. The use of wheat-based food discourses to regulate wider social and political anxieties is equally evident in the current dietetic moment. Remnants of

Grahamite sentiment can be read in the fears of wheat and gluten as a form of fatphobia and the social regulation of thinness as the ideal moral standard, one exercised through food choice.

Gluten aversions and the "new" fatphobia

As many in Western and North American contexts may recall, the US and Canadian governments of the 1980s actively pushed low-fat eating. Food guides, magazine covers, medical offices, and product labels all bore the same memorandum: choose low-fat products to mitigate your risk of "obesity" and cardiovascular disease. Guided largely by the work of Ancel Keys, an American physiologist who studied the effects of saturated fats on human health, this hypothesis was based on the claim that saturated fats derived from animal sources (red meats, egg yolks, lard, and high-fat dairy) carried the greatest risk to human cardiovascular health (CBC News, 2016a). Keys's claim was deduced from a seven-country study where he found that countries with high levels of heart disease consumed high levels of saturated fat while countries with low levels of heart disease consumed lower levels of what were quickly becoming known as "bad fats" (Taubes, 2008). Keys articulated what would become the low-fat craze of the next 15 to 20 years, which saw many North Americans diligently separating egg yolks from egg whites, purchasing skim or 1% dairy varieties over the full-fat equivalents, and spreading margarine instead of butter on toast. During this time, North Americans were also consuming more calories from wheat and grains as a response to formal dietary advice and presumably also filling the hunger left by a dearth of fat in their diets. By the late 1990s, Canadians consumed as much as 30% of their calories from grain and, at the height of the low-fat movement, were collectively gaining weight (CBC News, 2016a; Taubes, 2002, 2008).

Sceptics from a range of health and dietary professions started to question the demonization of dietary fat over refined starches and grains. Dr. Walter Willett, professor of epidemiology and nutrition and chair of the Department of Nutrition at the Harvard School of Public Health, articulated the false logic of the fat-free movement in the following crude yet to-the-point statement:

> The way you fatten up any animal is to put them in a pen so they can't run around, and then you feed them lots of grain. [...] We've basically created a North American feed lot where we have people exercising very little, eating lots of refined starches [...] And not surprisingly, we're getting fat.
>
> (CBC News, 2016a)

Similarly, the biochemist Dr. Barry Sears pointed out that where Keys went wrong was to blame fats and not insulin. He states that in essence we've been fighting the wrong enemy the last 15 years. [...] The enemy has been excess production of insulin [and one way] to produce excess levels of insulin in your body [...] is to eat too many fat-free carbohydrates.

(CBC News, 2016b)

Armed with this revised dietary advice, many North Americans swapped out bagels for eggs and bacon in their grocery carts, some even going as far as to spread butter not on their toast but on the bacon itself, as Dr. Jay Wortman professes in support of the keto diet (CBC Radio, 2019). What was once a social disdain for dietary fats became a fervent aversion to carbohydrates. While the specific food advice was repackaged from cholesterol- to wheat-free, the underlying dietetic message is both consistent and misguided: that being fat and being sick are synonymous.

In returning to the work of the grain-free gurus of the late 20th and early 21st centuries, the conflations between gluten consumption, being overweight, and ill health are overt. The subscript to *Dr. Atkins' New Diet Revolution* reads, "An indispensable guide to weight loss, weight maintenance, good health, and disease prevention" (p. i). Atkins (1992/2002) goes on to hail his low-carb diet as a way to "lose weight!", "increase energy!", and "look great!" (p. 3). In addition to weight loss, his low-carb diet is touted as "a revolutionary method for living a long, healthy life" (Atkins, 1992/2002, p. 4). Similarly, the subscript to Cordain's (2002) *Paleo Diet* is "to lose weight and get healthy" (p. i). And, drawing on negative fat stigma, is Davis's (2011) book title, *Wheat Belly: Lose the Wheat, Lose the Weight, and Find Your Path Back to Health*. In step with Atkins and Cordain, Davis pairs fatness with illness and does so through negative, stigmatizing associations of being overweight. In his words:

[W]heat's impact of the waistline is its most visible and defining characteristic, an outward expression of the grotesque distortions humans experience with this grain. A wheat belly represents the accumulation of fat that results from years of consuming foods that trigger insulin. [...] While some people store fat in their buttocks and thighs, most people collect ungainly fat around the middle.

(Davis, 2011, p. 4)

Davis's use of the word "belly" in the title of his book is pointed, as are his associations of stomach fat as "grotesque" and "ungainly". As many were avidly promulgating the sins of carbohydrates, fat stigma circulated under the seemingly cleansed, nutricentric category of gluten- and carb-free eating. Articulated in much of the grain-free claims popularized by Atkins, Cordain, and Davis, is the idea that being overweight is undesirable and tantamount to illness, neither of which have been scientifically confirmed.

Critical food, health, and fat studies scholars have pulled apart the many false conflations between health and weight that circulate in dominant dietetic discourse, including wheat-free claims. As Rich and Evans (2005) note, the epidemiological connections between fatness and sickness are so commonplace that we often overlook the simple truth that "one may be fat and healthy" or thin and unhealthy (p. 8). Many of the claims of "obesity" and ill health are based on body mass index (BMI) measurements, which reductively equate health with weight and height—an equation that reveals very little about a person's health (Rich & Evans, 2005). More telling, non-weight-based measurements of health may assess factors such as fitness capacity, sleep patterns, and stress levels. Kathleen Lebesco (2010) also undermines empirical claims that position "obesity" as a major cause of mortality and pathology. Citing recent epidemiological literature, she points out that "the vast majority of people labeled 'overweight' and 'obese' according to current definitions do not face any meaningful increased risk for early death" (Lebesco, 2010, p. 74). In some cases, being overweight can also be linked to a *reduced* risk in mortality, diabetes, and cardiovascular disease—a trend we do not hear much about in mainstream food or dietetic wisdom (Rochefort, Senchuk, Brady, & Gingras, 2016). Lucy Aphramor (2005) argues that weight-centred health frameworks are salutogenic, which means they over-focus on individual factors that may *support* human health as opposed to the social and systemic factors that *cause* disease. The hyper-focus on wheat- and gluten-free eating, popularized by the grain-free movements of the last 30 years, is equally salutogenic, ignoring structural causes of ill-health.

As Lebesco (2004, 2010), Coveney (2006), and Gard and Wright (2005) have extensively argued, the fatphobia, fat stigma, and weight shaming that circulates through diet-based discourses moralize individual choice, without questioning the social or structural access to health and healthy eating, a parallel in the discourses of wheat-free eating. As was the case in early health-based wheat discourses, the current rhetoric of wheat-free eating normalizes and even condones public scorn against fat, fat people, and fat bodies, which is a mode of governing and disciplining populations through the further regulation of socially marginalized others. As Lebesco (2004) sums up, being fat

> marks one as a failure at attaining citizenship in the dominant socioeconomic class. In wealthy nations [...] social constraints against 'vulgar' fatness are deployed by the dominant classes who are more willing and most able to produce the bodily forms of highest value as their formation requires investments of spare time and money.
>
> (p. 58)

Grain-free eating is expensive and time-consuming and upholds 21st-century nation-building discourses that sometimes subtly and sometimes overtly dictate which bodies deserve to live and which deserve to die. The social purity movement introduced by Graham in the early 19th century is alive and well and continues to target the poor, people of colour, and women through grain-based health claims. While Grahamite wheat discourses promoted social purity through wheat consumption, current grain-free eating dictums promote it through wheat aversion. The social regulatory rhetoric persists, even if the specific health-based advice about wheat is divergent.

Part of why the fat-free movement had such rhetorical success in the 1980s is because people equated the fats found in food with the accumulation of body fat (CBC News, 2016b). However, once the roles of dietary fats had been clarified, or at least complicated by the role of insulin, refined carbohydrates stepped in as the dominant dietetic demon of weight-loss efforts. In both public and medical discourse, gluten-, wheat-, and grain-free has replaced fat-free as the socially acceptable, nutricentric practice of weight-loss. For celebrities such as Gwyneth Paltrow, Oprah Winfrey, and Dr. Oz, as well as their followers, grain-free eating was the new dietetic solution to the socially constructed problem of "obesity" as well as cancer, heart disease, and diabetes. Lebesco (2004) asserts that "one might interpret the reinvigorated efforts to wipe out obesity as a sort of modern-day eugenics campaign: keep us pure by keeping us free of the 'maladies' experienced by cultural Others" (p. 59). In this sense, despite the debate about whether it was saturated fats or refined carbohydrates that were chiefly responsible for our collective BMIs increasing (rather than the problems with BMI measurements themselves), each side of the pro- and anti-wheat debate circulate the same underlying skewed logic: that it is our individual eating habits that are the problem and not the social, political, and economic conditions that produce ill-health or the social discourses that regulate healthy eating.

Wheat-free eating: a modern solution to a modern problem

Just as store-bought white bread was once configured as a "modern solution to a modern problem" (Bobrow-Strain, 2013, p. 265), wheat-free eating is yet another instantiation of individualized solutions to structural problems. A major issue with today's medicalized discourse of wheat-free eating is also a problem with dominant, contemporary discourses of health, nutrition, and dieting culture. As sociologists have long maintained, health and nutrition are problems of equity, not morality, and in the many current preoccupations over wheat-free eating, the structural conditions that give rise to social disparity are overlooked. Health and nutrition have come to be seen as a "'bio-problem', [and] a problem of life and living", not as a political or economic position (Coveney, 2006, p. 92). With the central responsibility for health placed on what we put in our stomachs, "nutrition discourses are salvation oriented, ascetic by nature, and [highly] individualized" (Coveney, 2006, p. 100). In dominant nutrition-based discourses, we blame "bad food choices" over structural poverty, "indolence" over systemic racism, and "bad genes" over gendered differences in spare time.

In her book, *Weighing In*, Julie Guthman (2011) documents how poverty, a lack of access to housing, employment, and food security, as well as other frightening logics of late capitalism are the obvious structural obstacles to health. The problem of obesity for Guthman (2011), and the problem of wheat-free eating for me, is that they are "artifact[s] of particular ways of measuring, studying, and redressing the phenomenon so that existing assumptions about its causes, consequences, and solutions are built into existing efforts to assess it independently" from the structural work of social justice (p. 23). Positioning fatness as the cause of the country's top killers and grain-free eating as the solution to it is equivalent to the 19th-century strategy of building empire through the perceived problem of racial degeneracy: it falsely constructs a social problem and presents a solution that obscures how the problem was constructed in the first place. Just as "racial degeneracy" was once thought to be "solved" though colonization, "obesity" is purportedly "cured" one grain-free meal at a time.

Historically, anti-wheat arguments merged with early nationalist racism, classism, and misogyny; currently, our collective gluten-phobia reinforces and normalizes an exclusionary, thin body politic and a further individualized eating discourse. What is upheld when this dietetic framing is accepted are the many risks associated with dieting, including but not limited to size discrimination and ill health through distorted relationships between food and health; the assumption that thinness and weight loss are of universal good; that losing weight will cure or prevent life-threatening diseases; that diets are effective for weight loss; and that dieting and weight loss is good for people's health (Aphramor, 2005; Rich & Evans, 2005). Wheat-free eating is a relatively new dietary fad (with the exception of the paleolithic diet some 200,000 years ago) and the long-term health consequences of strict wheat- and gluten-free eating are not yet known (CBC Radio, 2019). Gluten-free products are a $973 million industry in the US alone and represent a booming percentage of food commerce (Moore, 2017). In the current post-truth context where plural and conflicting truths about wheat's healthfulness circulate, let's critically question the centrality and history of wheat in the North American diet, but let's also equally question the discursive role of grain-free fads in the current health and illness landscapes.

In articulating a stance against singular dietetic truths, I aim to move beyond circular discussions that polarize pro- and anti-wheat positions in order to better examine the wider, structural effects that mitigate the social framings of wheat's dietetic discourse. The history of wheat and its perceived impacts on human health tell us more about the regulation of social marginalization and otherness than it does any simplistic dietary facts about the staple grain. Shifting food facts about wheat have been shaped alongside cultural anxieties that go far beyond the grain itself. The social anxieties of the 19th century circulated around perceived class, race, and gender impurities. Contemporarily, they circulate around fatphobia and the emphasis on individual food choice as cultural scapegoats for much more insidious social issues—issues of grotesque wealth disparity and unwieldly political control. In focusing too intently on wheat's biophysiological, nutricentric properties, dominant nutritional discourses side-step the cultural contexts in and through which arguments for and against the grain come to be culturally ingested. Its farinaceous properties reveal sticky biocultural entanglements that continue to individualize and moralize food choice within shifting social constructs.

Notes

1 Because of the overlapping qualities of wheat, gluten, and grain, throughout this chapter I'll use the three terms interchangeably, except when a distinction between them is required. I have chosen to use "wheat" in the chapter over "grain" or "gluten" (i.e., the collective term used to describe storage proteins in grains) because wheat is the most common and recognized of the three terms.
2 Known colloquially as the "keto" diet, the ketogenic diet is a high-fat, high-protein, low-carb diet initially used to treat epilepsy in the 1920s, but is more widely used now for weight loss and reversal of type 2 diabetes. Its name stems from the process of ketosis, where the body is starved of sugars and, as a result, burns stored fats (CBC Radio, 2019).
3 Celiac disease is a chronic autoimmune condition in the upper small intestine where the body lacks the ability to tolerate and digest gluten. The conditi on has been recorded since Antiquity, but was not associated with wheat until Dr. Dicke, a Dutch pediatrician, pointed out the decline in the disease during World War II (Bjørnstad, 2012).
4 Amy Kaplan (2002) writes about the paradox of imperial domesticity, where home became an engine of nationalist expansion and women's domestic labour contributed to the colonizing work of empire.

References

Agriculture and Agri-Food Canada. (2014, April). *"Gluten-free" claims in the marketplace* [PDF]. Retrieved from http://www.agr.gc.ca/resources/prod/doc/pdf/free_claims_gluten_sans_allegations2014-eng.pdf

Angel. (2016). Eye-opening! [comment 8]. Message posted to https://www.chapters.indigo.ca/en-ca/books/wheat-belly/9781443412735-item.html

Aphramor, L. (2005). Is a weight-centred health framework salutogenic? Some thoughts on unhinging certain dietary ideologies. *Social Theory and Health, 3*(4), 315–340.

Atkins, R. C. (1981). *Dr. Atkins' diet revolution.* London, UK: Bantam Press.

Atkins, R. C. (2002). *Dr. Atkins' new diet revolution.* New York, NY: M. Evans & Company Inc. (Original work published in 1992).

Beatrice. (2011, October 6). Consequences of genetic modification revealed! Message posted to https://www.amazon.ca/gp/review/RFGA7JZQWFW7T?ref=pf_vv_at_pdctrvw_srp

Bjørnstad, Å. (2012). *Our daily bread: A history of cereals.* Latvia: Vidarforlaget.

Bjørnstad, Å. (2016). Wheat: Its role in social and cultural life. In A. P. Bonjeam, W. J. Angus, & M. Van Ginkel (Eds.), *The world wheat book: The history of wheat breeding* (pp. 1–30) Cachan, FR: Lavoisier.

Bobrow-Strain, A. (2007). Kills a body in twelve ways: Bread fear and the politics of "what to eat?" *Gastronomica: The Journal of Food and Culture, 7*(2), 45–52. doi: 10.1525/gfc.2007.7.3.45

Bobrow-Strain, A. (2012). *White bread: A social the history of the store-bought loaf.* Boston, MA: Beacon Press.

Bobrow-Strain, A. (2013). White bread biopolitics: Purity, health, and the triumph of industrial baking. In R. Slocum & A. Saldana (Eds.), *Geographies of race and food: Fields, bodies, markets* (pp. 265–290). Burlington, VT: Ashgate.

Canadian Cancer Society. (2018). Eat well. Retrieved from http://www.cancer.ca/en/prevention-and-screening/reduce-cancer-risk/make-healthy-choices/eat-well/antioxidants-and-phytochemicals/?region=ab

Canadian Clinicians for Therapeutic Nutrition. (2019). Canadians are sick and overweight. Retrieved from https://ccfortn.ca/why-we-exist/

Canadian Wheat. (2018). Clean. Consistent. Quality. 2018 crop in review. Retrieved from https://canadianwheat.ca/review/Canadian%20Wheat%202018%20Crop%20in%20Review_181210.pdf

CBC News. (2016a, December 23). Fat and sugar, Pt. 1. *Ideas.* Retrieved from https://www.cbc.ca/radio/ideas/fat-and-sugar-part-1-1.3635209

CBC News. (2016b, December 30). Fat and sugar, Pt. 2. *Ideas.* Retrieved from https://www.cbc.ca/radio/ideas/fat-and-sugar-part-2-1.3645937

CBC News. (2018, March 6). Baker's asthma: Edmonton study to examine long-term health hazards of flour dust. *CBC News.* Retrieved from http://www.cbc.ca/news/canada/edmonton/baker-breath-lung-health-research-edmonton-flour-1.4564446

CBC Radio. (2019, January 11). Doctors who champion low-carb, high-fat diets go against the grain. *White Coat, Black Art.* Retrieved from https://www.cbc.ca/radio/whitecoat/doctors-who-champion-low-carb-high-fat-diets-go-against-the-grain-1.5140630

Cochrane, A. (2013, June 21). Modern wheat a perfect, chronic poison, doctor says. *CBS News.* Retrieved from https://www.cbsnews.com/news/modern-wheat-a-perfect-chronic-poison-doctor-says/

Cordain, L. (2002). *The paleo diet: Lose weight and get healthy by eating the food you were designed to eat.* Hoboken, NJ: John Wiley & Sons.

Coveney, J. (2006). *Food, morals and meaning: The pleasure and anxiety of eating* (2nd ed.). New York, NY: Routledge.

Curtis, B. C. (2002). Wheat in the world: FAO plant production and protection series. In B. C. Curtis, S. Rajaram, & H. Gómez Macpherson (Eds.), *Bread wheat: Improvement and production.* Food and Agriculture Organization of the United Nations. Retrieved from http://www.fao.org/docrep/006/y4011e/y4011e00.htm#Contents

Davis, W. (2011). *Wheat belly: Lose the wheat, lose the weight, and find your path back to health.* Emmaus, PA: Rodale.

Eaton, E. (2013). *Growing resistance: Canadian farmers and a politics of genetically modified wheat.* Winnipeg, MB: University of Manitoba Press.

Eaton, S. B, & Konner, M. (1985). Paleolithic nutrition: A consideration of its nature and current implications. *The New England Journal of Medicine, 312*(5), 283–289.

Elder, J. H. (2008). The gluten-free, casein-free diet in autism: An overview with clinical implications. *Nutrition in Clinical Practice, 23*(6), 583–588. doi: 10.1177/0884533608326061

Fasano, A., Sapone, A., Zevallos, V., & Schuppan, D. (2015). Nonceliac gluten and wheat sensitivity. *Gastroenterology, 148*(6), 1195–1204. doi: 10.1053/j.gastro.2014.12.049

Gard, M., & Wright, J. (2005). *The obesity epidemic: Science, morality, and ideology.* New York, NY: Routledge.

Gardner, C. D., Kiazand, A., Alhassan, S. Kim, S., Stafford, R. S., Balise, R. R., Kraemer, H. C., & King, A. C. (2007). Comparison of the Atkins, Zone, Ornish, and LEARN diets for change in weight and related risk factors among overweight premenopausal women. *Journal of the American Medical Association, 297*(9), 969–977. doi: 10.1001/jama.297.9.969

Gilman, S. L. (2008). Sylvester Graham. In *Diets and dieting: A cultural encyclopedia* (pp. 120–121). New York, NY: Routledge.

Graham, S. (1837). *A treatise on bread and breadmaking.* Boston, MA: Light and Stearns.

Graham, S. (1838). *A lecture to young men on chastity.* Boston, MA: Light and Stearns.

Guthman, J. (2011). *Weighing in: Obesity, food justice, and the limits of capitalism.* Berkeley, CA: University of California Press.

Heart and Stroke Foundation of Canada. (2018). Whole grains. Retrieved from http://www.heartandstroke.ca/get-healthy/healthy-eating/whole-grains

Kaplan, A. (2002). Manifest domesticity. In C. N. Davis & J. Hatcher (Eds.), *No more separate spheres! A next wave American studies reader* (pp. 183–208). Durham, NC: Duke University Press.

Kaplan, S. L. (2006). *Good bread is back: The contemporary history of French bread, the way it is made, and the people who make it.* Durham, NC: Duke University Press.

Katz, D. L. (2003). Pandemic obesity and the contagion of nutritional nonsense. *Public Health Review, 31*(1), 33–44. Retrieved from https://publichealthreviews.biomedcentral.com/

Lagiou, P., Sandin, S., Lof, M., Trichopoulis, D., Adami, H., & Weirderpass, E. (2012). Low carbohydrate-high protein diet and incidence of cardiovascular diseases in Swedish women: prospective cohort study. *British Medical Journal, 345*(7864), 16. doi: 10.1136/bmj.e4026

Lebesco, K. (2004). *Revolting bodies?: The struggle to redefine fat identity.* Boston, MA: University of Massachusetts Press.

Lebesco, K. (2010). Fat panic and the new morality. In J. M. Metzl & A. K. Kirkland (Eds.), *against health: How health became the new morality* (pp. 72–82). New York, NY: New York University Press.

Levinovitz, A. (2015a). *The gluten lie: And other myths about what you eat.* New York, NY: Regan Arts.

Levinovitz, A. (2015b). The problem with David Perlmutter, the brain grain doctor. *The Cut.* Retrieved from https://www.thecut.com/2015/06/problem-with-the-grain-brain-doctor.html

Magnan, A. (2016). *When wheat was king: The rise and fall of the Canada-UK grain trade.* Vancouver, BC: UBC Press.

McNeil, S. (2010). Traditional diet helps beat diabetes, says doctor. *Alberta Sweetgrass, 17*(12), 12–13.

Moore, L. R. (2017). Food intolerant family: Gender and the maintenance of children's gluten-free diets. *Food, Culture & Society, 20*(3), 463–483.

Olubukola, A., English, P., & Pinkney, J. (2013). Systematic review and meta-analysis of different dietary approaches to the management of type 2 diabetes. *American Journal of Clinical Nutrition, 97*(3), 505–516. doi:10.3945/ajcn.112.042457

Ornish, D. (2015, March 23). The myth of high-protein diets. *The New York Times*. Retrieved from https://www.nytimes.com/2015/03/23/opinion/the-myth-of-high-protein-diets.html

Perlmutter, D. (2013). *Grain brain: The surprising truth about wheat, carbs, and sugar—Your brain's silent killers*. New York, NY: Little, Brown and Company.

Probyn, E. (2013). Afterword: Biocultural entanglements. In R. Slocum & A. Saldana (Eds.), *Geographies of race and food: Fields, bodies, markets* (pp. 331–334). Burlington, VT: Ashgate.

Reeves, A. (2017). A critique of the "Paleo Diet": Broader implications of a socio-cultural food practice. *Contingent Horizons: The York University Student Journal of Anthropology, 3*(1), 1–6. Retrieved from https://contingenthorizons.files.word-press.com/2018/11/ch31-1-6-reeves.pdf

Rich, E., & Evans, J. (2005). "Fat ethics": The obesity discourse and body politics. *Social Theory and Health, 3*(4), 341–358.

Rochefort, J. E., Senchuk, A., Brady, J., & Gingras, J. (2016). Spoon fed: Learning about "obesity" in dietetics. In W. Mitchinson, D. Macphail, & J. Ellison (Eds.), *Obesity in Canada: Historical and critical perspectives* (pp. 148–174). Toronto, ON: University of Toronto Press.

Shewry, P. R. (2009). Wheat. *Journal of Experimental Botany, 60*(6), 1537–1553. doi:10.1093/jxb/erp058

Shikes. (2012, November 4). A rude awakening. Message posted to https://www.amazon.ca/gp/review/R9MDDDVTFZ27C?ref=pf_vv_at_pdctrvw_srp

Shotwell, A. (2016). *Against purity: Living ethically in compromised times*. Minneapolis, MN: University of Minnesota Press.

Taubes, G. (2002, July 7). What if it's all been a big fat lie? *The New York Times Magazine*. Retrieved from https://www.nytimes.com/2002/07/07/magazine/what-if-it-s-all-been-a-big-fat-lie.html?pagewanted=all

Taubes, G. (2008). *Good calories, bad calories: Facts, carbs, and the controversial science of diet and health*. New York, NY: Anchor Books.

Tomes, N. (2000). The making of a germ panic, then and now. *American Journal of Public Health, 90*(2), 191–199. Retrieved from https://ajph.aphapublications.org/

Tompkins, K. W. (2009). Sylvester Graham's imperial dietetics. *Gastronomica: The Journal of Food and Culture, 9*(1), 50–60. doi:10.1525/gfc.2009.9.1.50

Tompkins, K. W. (2012). *Racial indigestion: Eating bodies in the 19th century*. New York, NY: New York University Press.

Voegtlin, W. (1975). *The stone age diet: Based on in-depth studies of human ecology and the diet of man*. New York, NY: Vantage Press.

World Atlas. (2019). Top wheat producing countries. Retrieved from https://www.worldatlas.com/articles/top-wheat-producing-countries.html

5 Meat
False divides between veganism and carnism

The unresolved, heated debates about meat-eating and meat-avoiding in contemporary Western culture are familiar to many along the vegan, carnist, vegetarian, pescatarian, and flexitarian spectrums. Sifting through online and social media or following the latest health advice, debates about why, how often, and in what quantities humans should be consuming meat are difficult to avoid.[1] Meat-eaters often defend the human taste for animal flesh as primal and necessary for health, while vegans and vegetarians hold meat-eating in contempt for illness, the suffering of non-human animals, and environmental demise caused. Other emotional and utilitarian attachments to meat-eating and meat-avoiding abound, including those that tie to community, culture and tradition, family, religion, and social responsibility in an age of climate change. While meat still holds a prominent place in many North American restaurant menus and in the centre of many Western-based holiday meals, there is also greater availability of meat-free options in grocery-store isles and fast-food chains, the latter a telltale sign of meat avoidance having reached the cultural mainstream.

Early articulations of the benefits of vegetarianism, at least in Western contexts, emerged from the Temperance Movement, a late 19th- and early 20th-century American health movement where meat-eating was equated to heating and energizing foods (Tompkins, 2012). More contemporary advocacy about the many benefits of plant-based eating have been popularized through documentaries such as *Forks Over Knives* (2011), *Cowspiracy* (2014), and *What the Health* (2017). Highlighting the work of Dr. T. Colin Campbell's acclaimed, though controversial, "China Study",[2] *Forks Over Knives* (2011) and *What the Health* (2017) map the negative health impacts of animal-based foods on cardiovascular disease, cancer, and other chronic diseases, advocating that a plant-based diet can not only prevent but also reverse some of the effects of these illnesses. *Cowspiracy* (2014) explores the impact of animal agriculture on the environment and argues that the industry is a major propeller of global warming, water use, and deforestation. The films have been hailed by some as "eye-opening", "life-changing", and "important" and by other consumers as "vegan propaganda", "pseudoscientific", and representing "cherry-picked" data (Burrell, 2017; IMBd, n.d.;

Penner, 2017). The combination of ideological dogma and the deep-seated emotional bonds either for or against meat explain, in part, the attachments to the anti- and pro-meat divides. But popular documentaries and fringe health advocates are not the only ones warning against the dangers of meat.

In 2014, the World Health Organization (WHO) made international headlines by classifying processed meats such as bacon, hotdogs, and salami as a Group 1 carcinogen, meaning there is strong evidence that they cause cancer (Cancer Council, 2019; WHO, 2015). Red meats, including beef, pork, and lamb, are listed as a Group 2A carcinogen, meaning there are probable causes between their consumption and cancer (Simon, 2015; WHO, 2015). Also in 2014, the revisions to Brazil's Food Guide advocated for "minimally processed foods, in great variety, and mainly of plant origin" (Food and Agriculture Organization of the United Nations, 2019). Hailed by many health advocates as an international leader in food guide policy, Canada eventually followed Brazil's lead. The 2019 revisions to Canada's Food Guide took an equally strong stance in favour of plant-based proteins. Much to the Canadian beef, pork, and chicken industries' disproval, Health Canada urged Canadians to limit foods high in saturated fats and to choose plant-based proteins such as tofu, beans, nuts, and seeds where possible. Canada's 2019 Food Guide does not take a categorical stance *against* meat products or meat-eating. Yet, compared to historical editions, meat products are noticeably decentralized, both in terms of quantity and in the shift in language from "meat and alternatives" to "protein foods" (Health Canada, 2019a).

In response to the WHO statement on the dangers of processed meat consumption, online commenters were quick to speak up with statements such as, "I'd rather die than give up bacon", "It's been nice knowing you all", and "At least I'll die happy" (ABC News, 2015). Flippancy about death aside, the refusal to accept authoritative health advice in favour of one's personal convictions is characteristic of post-truth food politics. Equally, some vegans also go to extreme lengths to defend their eating habits. In a study on vegan-sexuality, defined by Potts and Parry (2010, 2014) as a lifestyle preference where vegans engage in sexual relations only with other vegans, one respondent states, "I couldn't think of kissing lips that allow dead animal pieces to pass between them" (quoted in Potts & Parry, 2010, p. 54). Another vegan-sexual similarly contends, "I wouldn't want to get close to someone in a physical sense if their body was derived from meat" (quoted in Potts & Parry, 2010, p. 54). Reproducing a strict, substance-based divide between veganism and carnism, vegan-sexuals conflate food politics with sexual orientation. Food choice, for both vegans and carnists, is much more complicated than singular truth-based and health-based concerns, and yet, often re-inscribes these divisions. Responses such as these speak to the dominance of food binaries based entirely on the material substance of food, rather than attitudes that consider multiple axes of social, political, and ethical eating—a point I will return to later in this chapter.

Despite the contempt for anti-meat sentiments by some, Westerners are, for the most part, eating less meat than they used to but are still eating more than the global average. Meat consumption in the US is more than three times the global average (Daniel, Cross, Koebnick, & Sinha, 2010). Tony Weis (2013) documents that because of Intensive Livestock Operations, combined with increasing global populations, the global per capita meat consumption has been steadily increasingly since the 1960s. He notes that in 1961, just over three billion people ate an average of 23 kg of meat per year, compared to 2011 when seven billion people ate on average 43 kg of meat per year (Weis, 2013). According to Statistics Canada (2019), the total meat consumption (i.e., including all red and white meats, seafood, and offal) per person in Canada has decreased since the 1980s. Beef consumption in particular has been steadily declining since its peak in 1976. While leading the meat market until 2000, beef sales have fallen substantially in the last 20 years, as chicken and turkey sales have risen. Beef currently sits at a distant second to chicken and turkey, just ahead of pork, which sits number three as measured by boneless weight before spoilage and waste per person per year (Statistics Canada, 2019). Chicken sales have increased in part because of marketing efforts that promoted chicken as a healthier, more convenient option to beef; its milder taste, more palatable to kids; and production efforts such rotisserie roasts and frozen chicken breasts, which make it more readily available at grocery stores (Dixon, 2008). In addition to the shift away from beef, Franklin (1999) and Palomo-Vélez, Tybur, and van Vugt (2018) speculate that the relative decline in North American meat consumption is a result of shifting health messages, a wider availability of vegetarian options, a growing awareness of environmental and ethical concerns around meat-eating, aging populations, and meat-based health scares, such as bovine spongiform encephalopathy (BSE), known informally as "mad cow" disease.

With decreasing sales, changing public attitudes concerning meat consumption, and recent shifts in government- and health agency-based dietary advice concerning meat intake, it is not surprising that meat industries are tactically shifting much of their consumer-directed messaging. For example, the Canadian Meat Council (2018) recently hailed meat a "superfood" that "strengthens the immune system" and "supports antioxidant production". Countering the popular claim that meat and vegetable-based protein sources are equivalent and interchangeable, they go on to compare various meat and plant-based food sources, concluding that "nothing is the same as meat" and "that imitation is the highest form of flattery", the latter of course directed at the growing market share of plant-based meat alternatives. The Alberta Beef Producers (2018) are also pointedly responding to the many fears concerning meat production, hormone, and antibiotic use. As stated on their website, they "work very closely with Animal Farm Care Animal Act and other organizations to exceed animal care standards"; that "hormones are a natural part of the beef", and that they "follow strict processes [for antibiotic use], including mandatory withdrawal times and residue testing to ensure that beef is antibiotic free". Attempting to reverse the label of tainted or unhealthy beef, the Alberta Beef Producers (2018) also draw on emotional connections by reminding consumers of the ways in which "beef is woven into the fabric of our lives" through "memories over slow-cooked ribs" and "perfectly grilled steaks". Further, they elicit historic and economic connections between beef and the local population by referring to the "long-standing cattle tradition" in Alberta and the reason "why in Alberta, we're #allforthebeef" (Alberta Beef Producers, 2018).

As Gwendolyn Blue (2008) argues, Alberta beef has long been associated with folk tradition and images of "wholesome cowboys and wild spaces" (p. 81). Not only is most of Canada's beef cattle raised and processed in Alberta, but it is such a strong provincial symbol, rooted in regional and national pride, appealing to collective identity and imagined communities, epitomized in "images of cattle grazing in the foothills of the Rockies under a clear [b]lue Alberta sky" (Gibson, 2003 quoted in Blue, 2008, p. 70). The depths of the regional attachments to Alberta beef were made even clearer in 2003 when a case of BSE—a neurodegenerative disease that affects cattle and can be transmitted to humans—was confirmed in a cow from Northern Alberta. Rather than questioning beef consumption, animal ethics, or the international integration, industrialization, and consolidation of the beef industry, Alberta ranchers, beef producers, the Alberta government, and the general public rallied together to support a fledgling industry. Concerted efforts were made to regenerate local industry through various government funds and fundraising events, including a 100 km barbeque and bumper stickers that read "I ♥ Alberta Beef" (Blue, 2008). In a show of what Theodore Platinga (2003) calls "mad cow nationalism", in the face of serious, life-threatening contagion what mattered more was the virility of the local economy and the sustainability of local beef production over any genuine critique of beef industry politics or meat-eating more broadly (quoted in Blue, 2008, p. 71).

As debates about meat continue in many contemporary contexts, it seems that for every argument on the pro-side, there is an equal and opposite argument on the anti-side and vice versa. Depending on who is consulted, meat is variably touted as a nutritious food source *and* one that contributes to disease risk, as an environmental hazard *and* one that is part of a healthy ecosystem, a food item that is both longstanding *and* extraneous to the human diet. I will first outline the pro- and anti-meat arguments in detail in order to unpack the seemingly contradictory terms of these debates. Next, I will analyze and critique these debates as being too didactic, and as creating and maintaining an either/or mentally instead of one that asks different questions about the healthfulness of meat in the human diet. By reframing the pro- and anti-meat debates, I aim to draw attention to the false health and ethical divides that get constructed along rigid vegan–carnist divides. I offer a reorientation away from circular discussions concerning meat-eating and meat-avoiding towards systemic and structural questions concerning health, food, illness, and the environment.

In what follows, I tease out five central themes in contemporary Western debates concerning meat-eating and meat aversion: meat's varied universality, its uniqueness as a food source, its carcinogenic and anti-carcinogenic properties, its perceived role in human weight loss and weight gain, and its environmental impacts and benefits. Rather than pitting the pro- and anti-meat debates against one another, as is often the case in mainstream and popular framings, I explore these five central themes to trace new questions of old debates and to move beyond narrow vegan–carnist divides. As I will expand in the final section of this chapter, by reframing the questions we ask of current meat debates, its social, ethical, and political context becomes more relevant than its substance—as has long been apparent in human meat consumption patterns.

Meat's varied universality

One of the existing debates in the cultural literature about meat as a food source is the degree to which humans have consumed meat historically and cross-culturally. Some sources liken us to our primate relatives, who survived largely—though not exclusively—on plant-based eating (Franklin, 1999). Other sources point to prominent ancient Greek and Roman philosophers who upheld the moral virtues of avoiding animals as food (Dombrowski, 2004). And others document longstanding vegetarian sects within societies, such as those of ancient and contemporary India, as well Buddhist or Jainist religious practices, but these tend to be the exceptions rather than the rule (Bulliet, 2007). Increasingly, there is archaeological, anthropological, and sociological evidence that humans of vast parts of the world have consumed meat, variously defined, for centuries (Daniel et al., 2010). In a study in the journal of *Meat Science*, N. J. Mann (2018) notes that contrary to the view that humans evolved largely as an herbivorist animal, archaeological evidence points to the central role of meat in the development of our species. Mann (2018) contends that as food availability changed from dense jungles to open grasslands, so did human food habits. While early humans were able to gather other sources, meat provided an optimal and more efficient source, and this holds true for exclusively carnist eating habits such as those in circumpolar regions that lack other food sources (Bulliet, 2007). In addition to widespread archeological evidence, Katherine O'Doherty-Jensen (2009) notes the anthropological and sociological patterns of global, regional, and local demand for meat over time, which "are neither recent nor local" (p. 1626). Most sources concur that humans are omnivores, though the degrees, quantities, and types of meat consumed have varied widely and have been dependent on food source availability, economic or class standing, religious or spiritual taboos, and social norms.

Anthropologists have documented that meat, more than any other edible substance, has been the subject of a wide range of food taboos and cultural regulation (Fessler & Navarette, 2003; Twigg, 1983). A mainstay in many human dietetic histories, the type and quantity of meat consumption remains highly socially regulated. Beef, for instance, as well as other forms domesticated cattle, are widely eaten in North and South America, Europe, Africa, and Australia and increasingly in parts of Asia (namely, Japan and China) as well. However, beef is also prohibited in many parts of India, where Hinduism upholds the cow as a sacred animal and where vegetarianism and the avoidance of flesh foods is associated with the higher ranks of the caste system (Beardsworth & Keil, 1997). Pork is also a staple for large portions of the world's population. It was a valued food for the Greeks and Romans, is common in present-day Europe and North America, and is widespread throughout Asia, particularly in China. Pork is also avoided by two of the world's major religions—Judaism and Islam—which see the animal as unclean and thus dangerous for human consumption. Chicken is widely consumed in large parts of the world, but avoided in parts of Sri Lanka and Tibet because of its association with filth and uncleanliness (Beardsworth & Keil, 1997). Fish is a fresh-water food source, but taboos exist on shellfish and "bottom feeders" like oysters, shrimp, clams, and mussels, again linked to their perceived dirtiness and therefore the potential contamination of human bodies (Cawthorn & Hoffman, 2016). While taboo in contemporary Western contexts, in times of economic, political, and/or environmental scarcity, rodents, marine mammals, non-human primates, and reptiles, are and have been food sources for various human populations (Cawthorn & Hoffman, 2016). The questions of which meats are consumed and avoided by humans and why are less about the substance of it and more about the cultural concerns, availability, and shifting health arguments tied to meat.

The more interesting critical question, for my purposes, is not whether meat has been consumed historically and cross-culturally, but rather the shifting, socially regulated arguments for and against meat consumption. Functionalist explanations for meat aversions typically encompass health and ecological rationales (Serpell, 2011; Twigg, 1983). One of the more frequently cited examples of a functional prohibition is the Jewish and Muslim avoidance of pork as a means of avoiding trichinosis—a parasitic disease common in pigs and other animals (Cawthorn & Hoffman, 2016; Simoons, 1994). Other functional meat taboos include the general avoidance of raw or uncooked meats and eggs because of their ability to render the human body ill. Fessler and Navarette (2003) also point to a possible functional explanation for the Hindu avoidance of beef—doing so protects cow from overexploitation, as they otherwise provide a stable source of milk for consumption, manure for farming, and some farm labour, all of which contribute to local food security, ecology, and economy. While functionalist explanations account for the explanation of some meat prohibitions, it is symbolic associations that typically dominate a culture's avoidance of meat, which are equally applicable to contemporary discussions of meat avoidance.

As the foundational work of Mary Douglas (1966/2003) has shown, symbolic rules around food and animal taboos frequently revolve around socially constructed ideals of purity and pollination and the connected perception that the animal in question that is either "sacred" or "profane". In almost all traditional tribal cultures, certain animals are revered as symbols of power, guardian spirits, and/or the residents of ancestors (i.e., totems). At times, the killing of the species is strictly prohibited. At other times, killing certain totem animals is sought out as a means of acquiring the animal's strength, virility, and power (Cawthorn & Hoffman, 2016). For example, for some tribes in East Africa, the flesh of a lion or a leopard is consumed by hunters in the hope that they will be imbued with the courageous and fierce characteristics of these animals. Relatedly, the meat, bone, blood, and body parts of various wild species are routinely used in traditional Chinese medicine due to their purported curative, aphrodisiac, and/or status-promoting properties (Cawthorn & Hoffman, 2016). Just as symbolic associations promote the consumptions of some animals, they also proscribe the consumptions of others. Margo DeMello (2012) notes that scavenger animals like vultures and rodents are frequently avoided as food sources due to their association with filth, disease, and death. Equally, positive symbolic links to dogs, cats, and horses as companion species, as well as to primates as human-like, typically render these animals inedible in Western contexts (Bulliet, 2007).

The functionalist and symbolic explanations that have historically and cross-culturally regulated meat consumption continue to circulate in contemporary Western debates about meat. Detailed in the following sections, functionalist arguments uphold links to the local economy, human health, and environmental ecology. Symbolic arguments maintain links to animal rights, the dangers of agro-industrial meat production, and constructed notions of "clean" or "pure" eating practices. The substance of meat and the associated debates concerning meat-eating and meat-avoiding are both socially bound and culturally regulated by the ways we think about and talk about these debates. I explore meat's uniqueness as a food source, its carcinogenic and anti-carcinogenic properties, its role in weight loss and weight gain, and its environmental hazards and aids to tease out the variations, contradictions, and plural truths in meat's contemporary cultural assessments.

Meat's uniqueness

The term "meat" in the English language signals cultural prominence as used to refer to the essence or most important feature, as in "the meat of an argument". Meat as essence or substance is used in symbolic contrast to the Western colloquial use of the term "vegetable" (or verb "to vegetate"), which implies a passive and inert existence (Beardsworth & Keil, 1997). The latter distinction is curious given the many health benefits of vegetables as well as the recent trend of certain vegetables being labelled superfoods because of their dense nutrient profiles. The socially constructed, inferior status of vegetables in many Western contexts is further evidenced by the central place of meat in most meal formats, with vegetables playing a secondary role as "the sides", especially pronounced in holiday meals and feasts (O'Doherty-Jensen, 2009). Meat's prominence is also indicated by the historical European association of meat with wealth, when royalty was one of the few groups who could afford to consistently eat meat. This pattern holds true today when looking at global meat consumption (Franklin, 1999). In addition to its uniqueness and social prominence, some discourses about meat also promote a kind of biochemical exceptionalism.

Confusion exists both within the literature and among the dietitians I interviewed for this research about meat's biochemical uniqueness. One of the dominant narratives about it, especially red meat, is that it provides micro- and macronutrients not easily found in other food sources. As one interviewee explains, "I believe that scientifically, there are components in meat, as a food matrix, that are not found in other supplements or replacements" (Spencer) and that meat "is an excellent way to get a number of components absorbed, vitamins, it's a rich source of protein, and certain lipids" (Spencer). Another interviewee also attests to "the importance of meat in maintaining muscle mass as we age" (Kai) and that meat "does have advantages over plant-based proteins" (Kai). Similar statements are conferred in the existing scientific and dietetic literatures.

Adrian Franklin (1999) notes that meat is considered one of the most nutritionally complete food sources for humans because of the similarities between animal and human flesh, making it a perfect source for muscle growth and recovery. Pereira and Vicente (2013) concur that meat is a concentrated nutrient source that aids in health, disease prevention, and brain development and that "the role of meat, especially red meat, is unequivocal" (p. 587). They go on to detail that red meat provides around 25% of the recommended dietary intakes for riboflavin, niacin, vitamin B6, and close to two-thirds the daily requirement of vitamin B12 (Pereira & Vincente, 2013). Meat is also widely hailed as the best source of essential minerals, including zinc, selenium, phosphorous, and iron in bioavailable form, as well as vitamin B12 (Beardsworth & Keil, 1997; Mann, 2018). While many other food sources, including beans, peas, lentils, fruits, vegetables, and grains, also carry iron, the iron in these foods is not always efficiently absorbed (Dietitians of Canada, 2014). B12 is also not found in most vegetable sources, though Barnard and Kieswer (2004) claim that vitamin B12 can be procured through fortified cereals, soymilk, and meat analogues. Pereira and Vicente (2013) also claim that B12 can be found in some types of algae, though it is admittedly not the most accessible food source, especially for land-locked populations. The elusive B12 vitamin is found most readily in liver and other meat sources, including fish, but also in dairy, eggs, and some plant-based milks.

Beardsworth and Keil (1997) wonder whether meat's unique biochemical makeup explains why human cultures have long gone to great extents to procure it as a food source. As they point out, "In terms of yield per unit of land, livestock cannot compare in efficiency with vegetable products" (Beardsworth & Keil, 1997, p. 201). Likewise, given the current ethical, environmental, and economic considerations surrounding animal feedlots and slaughterhouses, the question of "why meat" is a critical one, especially in the face of a growing availability of plant-based alternatives. There is some speculation that meat's distinct nutrient profile satiates hunger differently than plant-based foods alone. Marvin Harris (1986) coins the term "meat hunger" to refer a kind of hunger or craving for meat-based foods. In her sociological analysis, O'Doherty-Jensen (2009) found that two-thirds of adults deem a hot meal without meat incomplete, though it was not specified whether this was in relation to tradition, taste, fullness, or some combination thereof.

Unsurprisingly, marketing efforts put forward by the meat industry are also quick to draw on the discourses of meat's exclusive biochemical makeup and to promote this uniqueness as the cause and effect of good health. The Canadian Meat Council (2018) states that "only animal products [...] naturally contain Vitamin B12, the most absorbable form of iron and contain complete protein with all the amino acids needed for good health". Similarly, in a short YouTube video on how "Beef Stacks Up", Alberta Beef Producers (2018) cite that 75 grams of beef equals 26 grams of protein, 2.5 milligrams of iron, and 1.8 micrograms vitamin B12. In the face of expanding non-animal protein options for consumers, the Canadian Meat Council (2018) is quick to claim that meat "is still the ideal source". As convincing as these arguments may be, they are not unilaterally expressed.

A handful of dietitians I spoke to echoed that we do not need meat to be healthy or to receive vital micro- and macronutrients. Jessica asserts,

> If we're talking about misconceptions, there are a lot of misconceptions that you need to eat meat in order to be healthy. I think that you can get everything that you need nutritionally and environmentally. I think we need to look beyond the scope of just our own body.

Bianca equally attests, "I don't think we actually need red meat to survive. I think that we can get the nutrients we need from it from other sources". Victoria likewise claims, "I have worked with enough people to come to believe that if their iron is low, if their B12s are low, if they've got some amino acid imbalance, it can be corrected with plants". In a similar expression, Ray advises clients that "some meat in the diet is healthy", but that he does not "think it's entirely necessary". These professional opinions are corroborated by wider dietetic literature that supports that comprehensive vegan and vegetarian diets provide adequate nutrition and have demonstrated benefits in disease prevention and treatment (Craig & Mangels, 2009; Turner-McGrievy, Mendes, & Crimarco, 2017).

How can equally reputable sources say that meat-eating is necessary on one hand and on the other hand that it is extraneous? Moreover, if specialists like dietitians and academics cannot agree on meat's role in the human diet, then how is the general public to negotiate these conflicting truths? If we move away from singular food truths (i.e., statements that promote meat as unilaterally healthy or unhealthy), then it is possible that conflicting truths about meat can co-exist, especially in assessing that questions about meat's healthfulness reside in food context not food content. As the history and human cultures have shown, meat-eating varies widely. Its perceived healthfulness fluctuates based on a range of contextual factors. First, protein contents vary substantially between meat products. As Pereira and Vicente (2013) note, protein contents can be as high as 34.5% in chicken breast or as low as 12.3% in duck meat. Meat contents also range substantially based on fat content. Beef cuts range from 14% to 19% fat, while pork products can range between 8% and 28% fat (Pereira & Vicente, 2013). Second, cooking and feeding methods can also influence meat's nutricentric composition. Consider the differences between deep frying and steaming fish, and between grass- and grain-fed beef. Third, the absorption of meat's nutrients is affected by other dietary factors, including other vitamins and minerals. Scientists are just scratching the surface on other accessory factors such as epigenetics that affect the foods we consume (Biltekoff, Mudry, Kimura, Landecker, & Guthman, 2014). While meat may have some unique biochemical properties that are not shared by other food sources, the role of this uniqueness, how it interacts with different people with sometimes vastly different diets and social contexts is often overlooked in singular truth claims about meat's biochemical benefits.

Carcinogenic/antioxidant

Confusing and contradictory evidence also circulates about meat's role as a carcinogen. The WHO's classification of processed meats as a Group 1 carcinogen and red meats as a Group 2A carcinogen is corroborated by the International Agency for Research on Cancer (2018), the World Cancer Research Fund International (2018), Health Canada (2019b), and the Canadian Cancer Society (2019). Academic sources such as Daniel et al. (2010) and O'Doherty-Jensen (2009) also document that the evidence linking red and processed meats to colorectal cancer is convincing and suggestive for other cancers. Similarly, Nestle (2018) confirms that people who eat the highest quantities of meat display about 20% higher risk of colon and rectal cancers and seem to be at higher risk for cancers of the esophagus, liver, lung, and pancreas. One of my interview participants also upholds the risks of meat-eating and cancer. Brianna urges her to clients to avoid red meats if they can, based on the research she has seen from "blue zone populations"— ones who typically live longer and have lower rates of colorectal and other types of cancer. She goes on to say,

> A lot of people will come to me saying, 'Oh, I need to increase my iron, and my doctor or somebody recommended that I increase my red meats.' I'm like, 'Oh, would you like a side of colon cancer with your iron?'

The confusion around meat as a carcinogen stems from two arguments. The first is that animal foods do contain a number of bioactive agents that may positively affect health, including taurine and glutathione, which are antioxidants found in fish and meat (Mann, 2018). Red meat is also widely listed as a particularly rich source of conjugated linoleic acid, a family of trans fats that may have benefits in cancer prevention (Tree, 2018). But, in a classic example of nutricentrism, in this line of thinking, the nutrient components of meat are centralized over other aspects of the food, specifically quantities and cooking or curing methods. According to reputable sources, the average intake of red meat should not exceed 300 g per week and many westerners eat more than this (O'Doherty-Jensen, 2009; World Cancer Research Fund/American Institute of Cancer Research, 2018). In addition to this, cooking, preserving, and curing methods have significant impacts on meat's carcinogenic components. Specifically, cooking methods that involve high temperatures, such as frying, charring, or barbequing, as well as processing and curing methods like smoking or salting or the addition of chemical preservatives, lead to the formation of carcinogenic compounds in meat (Daniel et al., 2010). Cooking at high temperatures produces chemicals on the meat that cause mutation of its substance (Cross & Sinha, 2004; Mann, 2018). Avoiding direct flames and cutting away charred pieces may help reduce cancer risk, but neither of these are common in popular barbequing or grilling methods (Cross & Sinha, 2004). In returning to the question of whether meat causes cancer, meat itself—at least in small doses, steamed or baked—may not be the issue. As Franklin (1999) reminds us, "Apart from a few global locations, meat is not a staple. For most people, meat is an occasional food" (p. 145). The links between meat and cancer are much less about the substance of the food and much more about the contexts of meat-eating and meat-producing—they are reflective of situational, not universal, truths.

The second confusion that circulates around meat's role as a carcinogen stems from industry-funded efforts that cast doubt on the existing evidence on the known links between meat and cancer. Articulated on the Canadian Meat Council's (2018) website,

> There is no conclusive link between meat consumption and colon cancer. Some research has shown weak associations between overconsumption

of meat and colon cancer. However, no one food, including meat and prepared meat, has ever been found to have a cause and effect relationship with cancer.

The National Cattlemen's Beef Association also perpetuates uncertainty on the existing associations between meat and cancer; they state that "the available scientific evidence simply does not support a causal relationship between red or processed meat and any type of cancer" (quoted in Nestle, 2018, p. 64). Also questionable, both sites go on to exonerate high-temperature grilling and frying of meats, which as cited above, is known to increase cancer risk. That much of this information is readily accessible by the general public and may confirm what they wish to hear (i.e., that meat is safe and healthy) only adds to the uncertainty circulating. As Marion Nestle (2018) sums up, "Science funded by the meat industry argues that meat is nutritious, necessary, and safe. Independently funded scientists advise eating less meat. Take your pick" (p. 64). In a post-truth food culture, the contextual, situational truths of meat (and cancer) are often obscured in the face of nutricentric, decontextualized, and industry-filtered accounts. Similar patterns exist in the debates on meat's role in weight loss and weight gain.

Weight loss/weight gain

As discussed in Chapter 4, the paleo and keto diets have garnered support from scientific, medical, and lay communities as effective weight-loss programs (McNeil, 2010). According to the scientific literature, recent studies demonstrate that the majority of individuals who consume a high-protein, high-fat diet lose more weight during the first three to six months compared with to those who follow more conventional, low-fat, calorie-reduced approaches (Shilpa & Mohan, 2018). Ting, Dugré, Allan, and Lindblad (2018) similarly find that ketogenic diets help people lose approximately 2 kg more than those who follow low-fat diets over the course of a year. Moreno, Crujeiras, Bellido, Sajoux, and Casanueva (2016) likewise show the effectiveness of ketogenic diets for weight loss over a two-year period and detail that in addition to sustained weight loss, ketogenic diets also decrease levels of visceral fat mass (again compared to traditional low-calorie diets). These findings, combined with high-profile celebrity endorsements by the likes of Kourtney Kardashian and LeBron James, are convincing and sway popular opinion. Equally convincing, however, is evidence that suggests that vegan diets are also an effective weight loss technique.

Popularized in documentaries such as *Forks Over Knives* (2011) and *What the Health* (2017), the evidence of vegan and vegetarian diets on weight loss is mounting. In their meta-analysis, Huang, Huang, Hu, and Chavarro (2015) found that participants following these diets lost, on average, between 1.5 kg and 2.5 kg, respectively, more than those following non-vegetarian ones. Likewise, in a study comparing the effectiveness of vegetarian and

Mediterranean diets, the former was found more effective in reducing low-density lipoprotein cholesterol levels (Slomski, 2018). Celebrities such as Venus and Serena Williams and Zac Efron also tout the weight loss benefits of vegan eating. The prevailing rationale for why vegan and vegetarian diets are so effective for weight loss is the influx of vitamin-rich, low-calorie, low-fat foods. However, a closer analysis of the prevailing debate between meat-eating, meat-avoiding, and weight loss shows that this is misconstrued.

In both vegan and keto diets, the focus is on the content (not the context) of the diet and the prevailing wisdom, at least in dominant discourse, is that it is the food that is enabling weight loss. However, in both cases, weight loss may be the effect of overall calorie reduction, combined with portion control, as well as internal and external tracking mechanisms common to both diets. In short, the reported weight loss from these diets may have nothing to do with the specifics of the food itself. As Wendy Bennett and Lawrence Appel (2016) concur, "Because interventions in weight-loss trials are commonly multi-factoral [...] it is difficult to tease apart the effects of a specific diet from behavioural changes" (p. 9). Other factors that complicate unobscured results on meat-eating and weight loss, if we return to the myths of nutritional precision outlined in the introduction, include the difficulty of accurate reporting rates from participants in diet-based studies; how foods are prepared; and the effects of sleep, exercise, stress, epigenetics, hormone levels, life stages, medications, and gut bacteria on a person's metabolism. Moreover, many weight-loss studies are not longitudinal, therefore there is very little long-term evidence to support or deny the claims of meat-eating and meat-avoiding for weight-loss strategies.[3] In shifting the popular conversation of weight loss to include contextual factors, we may find more consistent food facts that reside outside vegan–carnist divides. Until then, the results that link weight loss with keto and vegan diets will continue to be mixed, caught up in the misguided search for singular, substance-based food facts.

Environmental hazard/aid

The other seemingly contradictory debate circulating about meat-eating is the degree to which it is an environmental hazard or aid. Alongside the broader counter-cultural movements of the 1970s and 1980s in North America, which took up interrelated issues of peace, anti-nuclear protest, environmental activism, and animal liberation, anti-meat sentiments became linked to ecological justice and this association has only strengthened in the last 50 years. As industrial farming practices have also intensified in the last 50 years, the "ecological hoofprint" of Intensive Livestock Operations (ILO) or Concentrated Animal Feedlot Operations (CAFO) has been extensively documented (Weis, 2013).[4] Even with "ag-gag" laws that criminalize the recording of animal rights and environmental abuses at agricultural facilities in the US, France, and Australia, the laundry list of meat's environmental toll is growing. Bruce Myers (2014) notes that animal products are responsible for roughly one-fifth of humanity's influence on global warming. The industrial livestock sector is particularly problematic, given its inefficient use of the land and water resources needed to feed, produce, and manufacture meat; the loss of biodiversity common to CAFOs; the large amounts of waste generated; as well as the pollution of land and waterways through the extensive CO_2 and greenhouse gases generated in production and distribution (Laestadius, Neff, Barry, & Frattaroli, 2013; Myers, 2014; O'Doherty-Jensen, 2009; Palomo-Vélez et al., 2018). While the food sector as a whole is a large contributor to global warming, especially during the rise of globalization, the meat industry is a notorious contributor. Recent studies estimate that an average carnist diet of 100 g of meat per day produces roughly 2.5 times more greenhouse gas emissions than an average vegetarian diet (Scarborough et al., 2014). The general consensus in the literature is that plant-based food production is more sustainable than meat and beef production on a number of environmental fronts (de Vries & de Boer, 2010; Eshel & Martin, 2006; Tilman & Clark, 2014). However, recent arguments concerning meat—and specifically beef production—as an environmental aid claim the contrary.

Responding to the *Lancet*'s recommendation to reduce human beef consumption by 90% (Willett et al., 2019) as well as the 2019 Canada Food Guide's stance on the merits of plant-based eating, Danielle Smith (2019) of the *Edmonton Journal* writes, "If you care about the planet, eat more beef". She goes on to vow to double her beef consumption in the name of saving the planet. Her reasons, as she details, is because she believes there is a concerted attack on the Canadian beef industry that could be "dangerous to our health" and "devastating to our environment". Without beef, she explains, we would lose essential vitamins and nutrients as well as biodiversity that cattle ranching can uniquely provide through carbon sequestering. Similarly, Isabelle Tree (2018) of *The Guardian* writes, "If you want to save the world, veganism isn't the answer". Touting the environmental benefits of grazing animals and what they give back to the earth, she (2018) notes that cow dung "feeds earthworms, bacteria, fungi, and invertebrates [which] is a vital process of ecosystem restoration, returning nutrients and structure to the soil". Some of what Smith (2019) and Tree (2018) articulate is supported by academic literature, with the notable caveat that beef's environmental benefits apply only to grazing farms, not feedlots. William Teague (2018) notes that "although ruminants, and cattle in particular, have been accused of a litany of damaging impacts on the global environment and human well-being, on deeper investigation ruminants provide a number of important benefits for humans and the environment" (p. 1519). Some of these benefits include maintaining healthy soils, minimizing soil erosion, sustaining clean water, as well as supporting plants, animals, and other organisms. Provenza, Kronberg, and Gregorini (2019) also contend that some pasture-finished beef products have markedly lower climate impacts and contribute positively to the health of soils, plants, and humans.

With arguments on both sides of the environmental hazard/environmental aid debate, it is understandable why consumers are confused about the roles of meat production and consumption in the context of the environment. Ironically, as much as these debates centre around the question of meat-eating and meat-avoiding, the answer to the question of meat's role in the environmental context is about the ecological and industrial conditions in which meat is raised, produced, sold, and consumed. While *some* beef production may tout the benefits espoused above, these environmental aids apply only to free-roaming, pasture-raised cows, who graze on photochemically rich mixtures of grasses, forbs, shrubs, and tress—not to CAFOs, where the majority of the Western beef supply is produced (Provenza et al., 2019). Not only are the health benefits of beef affected when livestock are fed monoculture pastures or high-grain rations in feedlots, but soil, ecosystem, and water qualities are also compromised. Teague (2018) also notes the important difference between grazing ecosystems and the current ranching methods used in beef production: the former is more sustainable for human and environmental health and plays an important role in countering the loss of soil carbon. Discussions about meat's role and the environment, similar to those of antioxidants and weight loss, cannot simplistically be divided along vegan–carnist lines, despite the dominant binary between them.

False food divides

Commonly pitted against each other, the focus on vegan–carnist divides obscures other relational and contextual-based concerns about food and eating in the contemporary moment—a moment marked by an ever-expanding global industrial food system; decreasing meat sales in North America; and growing, chronic health issues especially in marginalized populations. Once again, the focus on food substance in the prevailing meat debates encourages us to think about the structural issues of cancer, weight loss, and the environment as being solved by what we put in our stomachs. The impetus to view health solutions as being solved through dietary means is an extension of nutritionism ideology, as well as a convenient marketing tactic that decontextualizes food knowledge to defend its own economic interests, as I have also highlighted in the cases of dairy and wheat. The singular truth-based approach at the core of the many of current meat and vegan divides, merely perpetuates this thinking.

In addition to sidestepping some of the most pressing health-based concerns we are facing—those that tie to social, economic, and political structures—I also struggle to accept the simplistic ethical dichotomies constructed between animal-based and non-animal-based eating. Too often in a vegan–carnist binary, vegan foods come to be articulated as "clean", "ethical", and "cruelty-free" (in contrast to the dirty associations of the meat industry) without any consideration of non-substance-based, ethical concerns. If ethical, healthy, and/or environmentally conscious eating is measured solely on an animal/non-animal food divide, we negate systemic concerns such as the deplorable economic and environmental conditions of migrant workers; how large-scale industrial food production affects local people, environments, and ecosystems; and how growing, rearing, and cooking matter more than the food itself. In encouraging a move away from dominant food binaries, veganism-carnism central among them, I reorient a discussion of the content of food towards one that is more strongly situational and relational, framed within the increasingly compromised social, economic, ecologic, and political contexts we find ourselves in (Shotwell, 2016).

Lisa Heldke (2012) explains that a substance-based food ontology "makes it easier to regard things as Other—to reify and separate them from me" (p. 70). By centring on the substance of food, epitomized in vegan and carnist dieting patterns, we absolve ourselves of ethical responsibility for the many near and distant marginalized others involved in food production and consumption. We also risk resurrecting what Alexis Shotwell (2016) calls "defensive individualism"—the assumption that the self is somehow separate from the world, and yet, simultaneously, requires defence against it (p. 11). Defensive individualism is exemplified in substance-based food regimes where we are made responsible for our health. To circumvent the pitfalls of a nutriticentric, food-as-substance ontology, I draw on Heldke's (2012) relational food ontology to situate eating practices within the broader conditions in which we live. Heldke (2012) explains that she "quite regularly eschewed meat, [but] was painfully irresolute when it came time to reject the readily-available brands of chocolate that use enslaved children in their production" (p. 67). She then goes on to question why she dutifully holds up some iterations of ethical eating while routinely overlooking or sidestepping others: "How can we complicate or nuance the stories we tell about these foods? What violence has heretofore been invisible to them? What compassion has been occluded?" (Heldke, 2012 p. 70). In polemic binary thinking, dominant power relations are often overlooked and upheld.

Veganism and vegetarianism are, to a degree, relational food practices. For many, vegan and vegetarian eating practices place us, at times uncomfortably, in relation to the social, political, economic, gendered, and colonial relations of meat-eating practices and norms (Adams 1990/2010; Harper, 2010). A core feature of veganism and vegetarianism is the recognition that food connects us to systems and structures beyond ourselves: animal warfare; ecological relationships; environmental justice; and gendered, raced, and classed relations. As Kenneth Shapiro (2014) writes, veganism is a way of being and experiencing the world, associated with deep interconnections between self, animals, and nature. Similarly, Laura Wright (2015) upholds veganism as a "philosophy and way of living which [...] promotes the development and use animal-free alternatives for the benefits of humans, animals, and the environment" (p. 2). However, where veganism and vegetarianism fail, just as carnist ideologies also do, is by upholding a dogged ideological position that a food's substance is its most relevant ethical or health marker.

Veganism is premised on a substance-based ontology where only some foods are acceptable—committed, in Heldke's (2012) words, to the "thingness" of food (p. 78). Veganism advances a relational approach to food, but does so paradoxically by upholding a substance-based ontology (Overend, 2019). Vegan food that is industrially grown, via migrant and/or colonial labour practices, and at the expense of non-renewable resources (as seen in the case of massive soybean plantations) is not ethical, healthy, or environmentally sustainable, even if the food's substance is vegetable based (Overend, 2019). And the same critique applies to a range of other vegan foods, because the questions concerning ethical, healthy, and sustainable eating are much more complex than a rigid substance-based binary can enable. As Kim TallBear (2019) similarly contends,

> For me there is no good way to consume anything that we put in our body in the kind of society that we're living in. [...] The problem is our food system [and] all of the bodies that are used to prop up that system, whether they're human or nonhuman are being violated and exploited. [...] Most of us don't consume vegetables in a proper way either.
>
> (pp. 60–61)

In the compromised times we find ourselves in, notions of food purity are not the solution (Shotwell, 2016).

While dominant food discourse typically upholds and normalizes an animal versus plant-based food dichotomy, when asked whether or not they considered meat to be a healthy food item, the vast majority of my respondents answered with some version of "it depends". They spoke, as I have in this chapter, to factors such as the context of the rest of the diet, quantities, quality, types of meat, cooking and rearing methods, affordability, and personal preference. Emma articulates that "the more processed the meat is, I'd say the less healthy it is. But again, I'm not going to go black and white on someone and say, 'Absolutely not, you can't consume this and have a healthy diet'". Andrea urges her clients with cancer to avoid meat:

> I tell a lot of my clients who have cancer to avoid meat. Also, for somebody with a compromised digestive system, I would tell them to back off of meat. If you're going to eat meat, it should be organic and grass fed.

And Lindsay, another respondent, explained that she personally doesn't consume meat because of a history of heart disease in her family, but professionally, she understands that it's such "a personal choice" for many of the people she works with. Meat, and more broadly protein sources, is not a singular, immutable, or homogenous entities. Rather, they are fluid entities, based on cultural knowledge, historical context, familial upbringing, religious proscriptions, political convictions, food source availability, accessibility, quantity, quality, frequency, digestive capacity, economic standing, ability level, community connection, and myriad other factors beyond protein substance. Questions concerning human, animal, ecological, and environmental health also cannot be simplistically reduced to substance-based discussions. The longer we obsess over the false divides between meat and vegan options, the longer we ignore the structural food systems that are without a doubt affecting our health faster than any singular food substance—meat or bean.

Notes

1 I use the term "meat" in its broadest definition to refer to the substance of a non-human animal, except when a distinction between different types of meat are needed. I include in this definition fish (with scales) and seafood (without scales), which are technically an animal substance, but not commonly classified as meat, especially within Western protein hierarchies, which prioritize the muscle tissue of mammals and poultry (Beardsworth & Kiel, 1997).
2 Based on the China-Cornell-Oxford Project, the 20-year multilateral study examined mortality rates from 48 forms of cancer and other chronic diseases, concluding that counties with a high consumption of animal-based foods were more likely to have had higher death rates from "Western" diseases, while the opposite was true for counties that ate more plant foods over the same period (Campbell & Campbell, 2006).
3 Existing feminist literature on dieting further confirms that the vast majority of diets are unsuccessful in the long run and carry long-term health consequences (Aphramor, 2005; Rich & Evans, 2005).
4 An ILO or CAFO is defined as a feedlot with more than 1,000 animal units (1 animal = 1 unit). Some of the largest US cattle feedlots have upwards of 20,000 cows on one site (Andrews, n.d.).

References

ABC News. (2015). Bacon, sausages, ham and other processed meats are cancer-causing, red meat probably is too: WHO. Retrieved from https://www.abc.net.au/news/2015-10-27/processed-meats-cause-cancer-says-un-agency/6886882

Adams, C. J. (2010). *The sexual politics of meat: A feminist vegetarian critical theory.* New York, NY: Continuum Publishing Company. (Original work published in 1990).

Alberta Beef Producers. (2018). Incredible things surround beef. Retrieved from https://www.albertabeef.org/consumers/home

Andersen, K. (Co-director), & Kuhn, K. (Co-director). (2014). *Cowspiracy: The sustainability secret* [Motion picture]. Los Angeles, CA: Appian Way Productions.

Andersen, K. (Co-director), & Kuhn, K. (Co-director). (2017). *What the health* [Motion picture]. Santa Rosa, CA: A.U.M. Films & Media.

Andrews, R. (n.d.). Cattle feedlot: Behind the scenes. Retrieved August 22, 2019 from https://www.precisionnutrition.com/cattle-feedlot-visit

Aphramor, L. (2005). Is a weight-centred health framework salutogenic? Some thoughts on unhinging certain dietary ideologies. *Social Theory and Health, 3*(4), 315–340.

Barnard, N., & Kieswer, K. (2004). Vegetarianism: The healthy alternative. In S. F. Sapontzis (Ed.), *Food for thought: The debate over eating meat* (pp. 46–56), Amherst, NY: Prometheus Books.

Beardsworth, A., & Keil, T. (1997). *Sociology on the menu: An invitation the study of food and society.* New York, NY: Routledge.

Bennett, W. L., & Appel, L. J. (2016). Vegetarian diets for weight loss: How strong is the evidence? *Journal of General Internal Medicine, 31*(1), 9–10. doi:10.1007/s11606-015-3471-7

Biltekoff, C., Mudry, J., Kimura, A. H., Landecker, H., & Guthman, J. (2014). Interrogating moral and quantification discourses in nutritional knowledge. *Gastronomica: The Journal of Critical Food Studies, 14*(3), 17–26. doi:10.1525/gfc.2014.14.3.17

Blue, G. (2008). If it ain't Alberta, it ain't beef: Local food, regional identity, (inter)national politics. *Food, Culture & Society, 11*(1), 69–85. doi:10.2752/155280108X276168

Bulliet, R. W. (2007). *Hunters, herders, and hamburgers: The past and future of animal-human relations.* New York, NY: Columbia University Press.

Burrell, S. (2017). Nutritionist Susie Burrell reviews Netflix's new documentary *What the Health.* Retrieved from http://www.news.com.au/lifestyle/health/diet/nutritionist-susie-burrell-reviews-netflixs-new-documentary-what-the-health/news-story/6bf4535413ffa0d08c3d1379af5c85cc

Campbell, T. C., & Campbell, T. M. (2006). *The China study: The most comprehensive study of nutrition ever conducted and the startling implications diet, weight loss, and long-term health.* Dallas, TX: BenBella Books.

Canadian Cancer Society. (2019). Why you should limit red meat and avoid processed meat. Retrieved from https://www.cancer.ca/en/prevention-and-screening/reduce-cancer-risk/make-healthy-choices/eat-well/limit-red-meat-and-avoid-processed-meat/?region=bc

Canadian Meat Council. (2018). Nutrition. Retrieved from https://cmc-cvc.com/consumers/nutrition/

Cancer Council. (2019). Meat and cancer. Retrieved from https://www.cancercouncil.com.au/21639/cancer-prevention/diet-exercise/nutrition-diet/fruit-vegetables/meat-and-cancer/

Cawthorn, D. M., & Hoffman, L. C. (2016). Controversial cuisine: A global account of the demand, supply, and acceptance of "unconventional" and "exotic" meats. *Meat Science, 120*, 19–36. doi:10.1016/j.meatsci.2016.04.017

Craig, W. J., & Mangels, A. R. (2009). Position of the American Dietetic Association: Vegetarian diets. *Journal of the American Dietetic Association, 109*, 1266–1282.

Cross, A. J., & Sinha, R. (2004). Meat-related mutagens/carcinogens in the etiology of colorectal cancer. *Environmental and Molecular Mutagenesis, 44*(1), 44–55. doi:10.1002/em.20030

Daniel, C., Cross, A., Koebnick, C., & Sinha, R. (2010). Trends in meat consumption in the USA. *Public Health Nutrition, 14*(4), 575–583. doi:10.1017/S136898001000207

DeMello, M. (2012). *Animals and society: An introduction to human-animal studies.* New York, NY: Columbia University Press.

de Vries, M., & de Boer, I. J. M. (2010). Comparing environmental impacts for livestock products: A review of life cycle assessments. *Livestock Science, 128*(1–3), 1–11. doi:10.1016/j.livsci.2009.11.007

Dietitians of Canada. (2014). Food sources of iron. Retrieved from https://www.dietitians.ca/Downloads/Factsheets/Food-Sources-of-Iron.aspx

Dixon, J. (2008). Operating upstream and downstream: How supermarkets exercise power in the food system. In J. Germov & L. Williams (Eds.), *A sociology of food and nutrition: The social appetite* (3rd ed., pp. 100–124). South Melbourne, Australia: Oxford University Press.

Dombrowski, D. (2004). A very brief history of philosophical vegetarianism. In S. F. Sapontzis (Ed.), *Food for thought: The debate over eating meat* (pp. 22–35), Amherst, NY: Prometheus Books.

Douglas, M. (2003). *Purity and danger: An analysis of the concept of pollution and taboo.* New York, NY: Routledge. (Original work published in 1966).

Eshel, G., & Martin, P. A. (2006). Diet, energy, and global warming. *Earth Interactions, 10*(9), 1–17.

Fessler, D. M., & Navarette, C. D. (2003). Meat is good to taboo: Dietary proscriptions as a product of the interaction of psychological mechanisms and social processes. *Journal of Cognition and Culture, 3*(1), 1–40.

Food and Agriculture Organization of the United Nations. (2019). Food-based dietary guidelines Retrieved from http://www.fao.org/nutrition/education/food-based-dietary-guidelines/regions/countries/brazil/en/

Franklin, A. (1999). *Animals and modern culture: A sociology of human-animal relations in modernity.* Thousand Oaks, CA: Sage Publications.

Fulkerson, L. (Director). (2011). *Forks over knives* [Motion picture]. Santa Monica, CA: Monica Beach Media.

Harper, A. B. (2010). *Sistah vegan: Black female vegans speak on food, identity, health, and society.* New York, NY: Lantern Books.

Harris, M. (1986). *Good to eat: Riddles of food and culture.* London, UK: Allen & Unwin.

Health Canada. (2019a). *Canada's Food Guide.* Retrieved from https://food-guide.canada.ca/en/

Health Canada. (2019b). *Canada's dietary guidelines: Section 1—Foundation for healthy eating* (Catalogue number H164-231/2019E-PDF). Retrieved from https://food-guide.canada.ca/en/guidelines/section-1-foundation-for-healthy-eating/

Heldke, L. (2012). An alternative ontology of food: Beyond metaphysics. *Radical Philosophy Review, 15*(1), 67–88. doi:10.5840/radphilrev20121518

Huang, R. Y., Huang, C. C., Hu, F. B., & Chavarro, J. (2015). Vegetarian diets and weight reduction: A meta-analysis of randomized controlled trials. *Journal of General Internal Medicine, 31*(1), 109–116. doi:10.1007/s11606-015-3390-7

IMBd. (n.d.). *What the Health.* Retrieved July 21, 2017 from http://www.imdb.com/title/tt5541848/

International Agency for Research on Cancer. (2018). *IARC monographs on the evaluation of carcinogenic risks to humans: Red meat and processed meat* (Vol. 114). Lyon, France. Retrieved from https://monographs.iarc.fr/wp-content/uploads/2018/06/mono114.pdf

Laestadius, L. I., Neff, R. A., Barry, C. L., & Frattaroli, S. (2013). Meat consumption and climate change: The role of non-governmental organizations. *Climatic Change, 120*(1), 25–38. doi:10.1007/s10584-013-0807-3

Mann, N. J. (2018). A brief history of meat in the human diet and current health implications. *Meat Science, 144,* 169–179. doi:10.1016/j.meatsci.2018.06.008

McNeil, S. (2010). Traditional diet helps beat diabetes, says doctor. *Alberta Sweetgrass, 17*(12), 12–13.

Moreno, B., Crujeiras, A. B., Bellido, D., Sajoux, I., & Casanueva, F. F. (2016). Obesity treatment by very low-calorie-ketogenic diet at two years: Reduction in visceral fat and on the burden of disease. *Endocrine, 54*(3), 681–690. doi:10.1007/s12020-016-1050-2

Myers, B. (2014). Livestock's hoof print. *The Environmental Forum, 31*(2), 34–39.

Nestle, M. (2018). *Unsavory truth: How food companies skew the science of what we eat.* New York, NY: Basic Books.

O'Doherty-Jensen, K. (2009, August). *Sociological aspects of meat in meals: Cultural impacts and meal patterns.* Paper presented at the 55th International Congress of Meat Science and Technology (ICoMST) conference, Copenhagen, Denmark.

Overend, A. (2019). Is veganism a queer food strategy? In B. Parker, J. Brady, E. Power, & S. Belyea (Eds.), *Feminist food studies: Exploring intersectionality* (pp. 79–101). Toronto, ON: Canadian Scholars' Press.

Palomo-Vélez, G., Tybur, J. M., & van Vugt, M. (2018). Unsustainable, unhealthy, or disgusting? Comparing different persuasive messaging against meat consumption. *Journal of Environmental Psychology, 58,* 63–71. doi:10.1016/j.jenvp.2018.08.002

Penner, J. (2017). *What the health* review: The good, the bad, and the ugly. Retrieved from https://www.smartnutrition.ca/nutrition-2/what-the-health-review-good-bad-ugly/

Pereira, P. M., & Vicente, A. F. (2013). Meat nutritional composition and nutritive role in the human diet. *Meat Science, 93*(3), 586–592. doi:10.1016/j.meatsci.2012.09.018

Potts, A., & Parry, J. (2010). Vegan sexuality: Challenging heteronormative masculinity through meat-free sex. *Feminism and Psychology, 20*(1), 53–72. doi:10.1177/0959353509351181

Potts, A., & Parry, J. (2014). "Too sexy for your meat": Vegan sexuality and the intimate rejection of carnism. In J. Sorenson (Ed.), *Critical animal studies: Thinking the unthinkable* (pp. 234–250). Toronto, ON: Canadian Scholars' Press Inc.

Provenza, F. D., Kronberg, S. L., & Gregorini, P. (2019). Is grassfed meat and dairy better for human and environmental health? *Frontiers in Nutrition, 6*(article 26), 1–13. doi:10.3389/fnut.2019.00026

Rich, E., & Evans, J. (2005). "Fat ethics": The obesity discourse and body politics. *Social Theory and Health, 3*(4), 341–358.

Scarborough, P., Appleby, P. N., Mizdrak, A., Briggs, A. D. M., Travis, R. C., Bradbury, K. E., & Timothy, J. K. (2014). Dietary greenhouse gas emissions of meat-eaters, fish-eaters, vegetarians and vegans in the UK. *Climatic Change, 125*(2), 179–192. doi:10.1007/s10584-014-1169-1

Serpell, J. A. (2011). One man's meat: Further thoughts on the evolution of animal food taboos. Retrieved from https://nationalhumanitiescenter.org/on-the-human/2011/11/one-mans-meat/

Shapiro, K. J. (2014). "I am a vegetarian": Reflections on a way of being. *Society & Animals: Journal of Human-Animal Studies, 23*(2), 128–147.

Shilpa, J., & Mohan, V. (2018). Ketogenic diets: Boon or bane? *The Indian Journal of Medical Research, 148*(3), 251–253. doi:10.4103/ijmr.IJMR_1666_18

Shotwell, A. (2016). *Against purity: Living ethically in compromised times.* Minneapolis, MN: University of Minnesota Press.

Simon, S. (2015, October 26). World health organization says processed meat causes cancer. Retrieved from https://www.cancer.org/latest-news/world-health-organization-says-processed-meat-causes-cancer.html

Simoons, F. J. (1994). *Eat not this flesh: Food avoidances from prehistory to the present*. Madison, WI: University of Wisconsin Press.

Slomski, A. (2018, April 24). Vegetarian and Mediterranean diets effective for weight loss. *JAMA, 319*(16), 1649. doi:10.1001/jama.2018.4738

Smith, D. (2019, January 25). If you care about the planet, eat more beef. *The Calgary Herald*. Retrieved from https://calgaryherald.com/opinion/columnists/smith-if-you-care-about-the-planet-eat-more-beef

Statistics Canada. (2019). Food available in Canada (Table 32-10-0054-01). Ottawa, ON. Retrieved from https://www150.statcan.gc.ca/t1/tbl1/en/tv.action?pid=3210005401

TallBear, K. (2019). Being in relation. In S. King, S. R. Carey, I. Macquarrie, V. N. Millious, & E. M. Power (Eds.), *Messy eating: Conversations on animals as food* (pp. 54–67). New York, NY: Fordham University Press.

Teague, W. R. (2018). Forages and pastures symposium: Cover crops in livestock production: Whole-system approach: Managing grazing to restore soil health and farm livelihoods. *Journal of Animal Science, 96*(4), 1519–1530. doi:10.1093/jas/skx060

Tilman, D., & Clark, M. (2014). Global diets link environmental sustainability and human health. *Nature, 515*(7528), 518–522. doi:10.1038/nature13959

Ting, R., Dugré, N., Allan, M., & Lindblad, A. J. (2018). Ketogenic diet for weight loss. *Canadian Family Physician, 64*(12), 906–906. Retrieved from https://www.cfp.ca/content/cfp/64/12/906.full.pdf

Tompkins, K. W. (2012). *Racial indigestion: Eating bodies in the 19th century*. New York, NY: New York University Press.

Tree, I. (2018, August 25). If you want to save the world, veganism isn't the answer. *The Guardian*. Retrieved from https://www.theguardian.com/commentisfree/2018/aug/25/veganism-intensively-farmed-meat-dairy-soya-maize

Turner-McGrievy, G., Mendes, T., & Crimarco, A. (2017). A plant-based diet for overweight and obesity prevention and treatment. *Journal of Geriatric Cardiology, 14*(5), 369–374. doi:10.11909/j.issn.1671-5411.2017.05.002

Twigg, J. (1983). Vegetarianism and the meanings of meat. In A. Murcott (Ed.), *The sociology of food and eating: Essays in the sociological significance of food* (pp. 18–30). Aldershot, UK: Gover Publishing Limited.

Weis, T. (2013). *The ecological hoofprint: The global burden of industrial livestock*. New York, NY: Zed Books.

Willett, W., Rockström, J., Loken, B., Springmann, M., Lang, T., Vermeluen, S., … Murray, C. (2019). Food in the anthropocene: The EAT–Lancet Commission on healthy diets from sustainable food systems. *The Lancet, 393*(10170), 447–492. doi:10.1016/S0140-6736(18)31788-4

World Cancer Research Fund International/American Institute for Cancer Research. (2018). *Diet, nutrition, physical activity and colorectal cancer*. Retrieved from https://www.wcrf.org/sites/default/files/Colorectal-cancer-report.pdf

World Health Organization. (2015, October). Q and A on the carcinogenicity of the consumption of red meat and processed meat. Retrieved from https://www.who.int/features/qa/cancer-red-meat/en/

Wright, L. (2015). *The vegan studies project: Food, animals, and gender in the age of terror*. Athens, GA: University of Georgia Press.

Conclusions
The trouble with singular food truths

When presenting parts of this book at conferences, I often anticipated a question that never came. The question not asked, the one I eventually posed to myself, was, "What do we lose if we move away from the search for singular food truths"? The short answer is, "A narrow, nutricentric, misguided, hegemonic, and individualist understanding of the role of food in our lives". The longer answer follows here.

In my push to contextualize and complicate singular food truths, I do not deny that some empirical ones, such as those explaining the necessary role of micro- and macronutrients in human health, have their place on the plate of dietetic advice. However, as I have argued throughout this book, we have tipped too far towards singular food truths at the expense of other ways of understanding food and health. As Mayes and Thompson (2014) attest, the relatively short history of nutrition science has provided a number of successful interventions into deficiency diseases. Crucially, they also note that the deficiency diseases of early nutritional science are vastly different in context and causation than the non-communicable, chronic diseases currently affecting Western societies (Mayes & Thompson, 2014). While an empirical nutricentric understanding of food worked well to mitigate the deficiency diseases of yore, the same model does not equally apply to the many chronic health concerns of today. In short, the causation between food, cancer, heart disease, and/or diabetes is far from definitive or straightforward, since each of these illnesses is not only biologically more complex than historic vitamin deficiencies, but these illnesses are also contingent on the conditions of social inequality not accounted for by a nutricentric food lens. To borrow the words of novelist Chimamanda Adichie (2009), in a different context than originally intended, "The danger of a single story is not that it is untrue, but that it is incomplete". The danger of empirical food truths is not that they are untrue, but that they are incomplete stories about food and health and the relationship between the two.

Documented in the genealogy of historical food framings in Chapter 1, the singular food truths of nutritionist ideology emerged alongside the object-based classifications of Enlightenment thinking. Since the shift away from the holistic, humoural food systems of the Ancient and Renaissance periods and towards the discovery of micro- and macronutrients in the late 19th and early 20th centuries, North American and Western understandings of food have been overwhelmingly decontextualized. From quantified, caloric food measurements to nutricentric marketing claims that advertise magic-bullet solutions to systemic and structural problems of living, to the many false binaries that come to regulate our diet and eating patterns, singular food truths abound, despite their many contradictory claims and ineffectiveness against non-communicable, chronic diseases. In the seemingly insatiable search for singular food truths, foods cycle between good/bad, natural/altered, healthy/unhealthy, ethical/unethical highlighting the shifting and constructed contexts of the food items in question and their variable classificatory schemes. The prevailing food truths concerning Western staple foods such as dairy, wheat, and meat, are not single or homogenous. And thus, to understand and deconstruct their proposed health benefits, I offer a contextual, relational analysis not enabled by a singular, nutricentric approach. My hope in moving beyond the search for and maintenance of singular food truths is twofold: (1) a shift from the well-worn question of "what to eat" to "why we eat the way we do" raises new questions about the discourses of food, nutrition, and health in the contemporary moment beyond object-based, individualist food strategies; and (2) a less authoritative sense of eating beyond the proscriptive, constrained approaches synonymous with Western nutricentric food ideologies that assume there is only one way to eat and only one way to eat *well*.

Relational, contextual eating

As a result of the growing conundrums of living in advanced capitalist societies, many are looking for ways to make food consumption healthier and/or more sustainable. However, the focus on singular food truths, which typically revolves around the substance of food, still places the responsibility for structural change on individual consumers. The last 200 years of food science has encouraged a singular, empirical view of food, which has produced a series of contradictory, conflicting, and at times misguided understandings of nutrition. If we continue to ask "either/or" questions of food—questions epitomized in debates of "dairy or dairy-free?", "wheat or wheat-free?", "meat or meat-free?"—we will continue to generate polarized, conflicting food truths that fail to address systemic and structural issues. We will also fail to "understand this food-body relationship as something expansive, intricate, and diverse"—a far more urgent task given current ecological landscapes (Hayes-Conroy & Hayes-Conroy, 2013, p. 1). While it may be compelling to romanticize singular food truths, especially amidst the growing cacophony of dietary advice on chronic diseases—drinking a glass of milk a day will reduce the risk of osteoporosis, cutting out carbs is an effective weight-loss method, and eating vegan will reduce the risk of cancer—no singular food item or singular food truth is a magic-bullet solution to the complex problems of living we are faced with. Part of my goal in moving away from singular food truths is to reframe health outside of individual food choices—a framing too often at the forefront of healthy eating initiatives.

In his essay "What Is Health and How Do You Get It?", Richard Klein (2010) reminds readers that "when anything has been so [...] glorified that its value is taken for granted, it becomes rhetorically necessary to exaggerate the value of its opposite in order for skepticism to be heard" (p. 15). Singular food truths are so extensively circulated, widely assumed, and generally presupposed conditions for health that we often fail to see, understand, judge, define, discuss, and even eat food outside of the grip of a nutricentric ideology. While singular food truths circulate through a range of contemporary food discourses, such as those upheld by food guides and other public health documents, popular discourse, the growing marketing and advertising efforts that normalize them, they are arguably most typified through the reductive "nutritional facts" tables that adorn food packaging (see example Figure 6.1).

Nutrition Facts Valeur nutritive	
Per 1/2 cup (125 mL) / par 1/2 tasse (125 mL)	
Amount **Teneur**	**% Daily Value** **% valeur quotidienne**
Calories / Calories 110	
Fat / Lipides 1 g	**2 %**
Saturated / saturés 0 g + Trans / trans 0 g	**0 %**
Cholesterol / Cholestérol 0 mg	
Sodium / Sodium 15 mg	**0 %**
Carbohydrate / Glucides 18 g	**6 %**
Fibre / Fibres 6 g	**24 %**
Sugars / Sucres 0 g	
Protein / Protéines 7 g	
Vitamin A / Vitamine A	**0 %**
Vitamin C / Vitamine C	**0 %**
Calcium / Calcium	**6 %**
Iron / Fer	**15 %**

Figure 6.1 Sample nutrition facts table.
Source: Author's own photo.

Intended as a way to provide consumers information about food, nutrition fact tables also uphold and normalize the idea of singular food truths, reduced to numeric calorie intakes and quantified percentages of micro- and macronutrients, literally omitting contextual or relational truths from the analytic frame. As Jessica Mudry (2009) likewise explains, "The use of quantitative language to describe food is a failure. It distorts judgments about eating, discounts and other knowledge claims regarding health, and unfairly reimagines the nature of food" (p. 3). While my aim to devalue the singular truths so highly prized and often unquestioned in current food and health frameworks is new, I am not alone in exposing other ways to revalue and reevaluate food and eating. In *Dangerous Digestion*, E. Melanie DuPuis (2015) articulates an approach to food knowledge beyond the binaries of purity and danger so commonly at the forefront of American dietary advice. She documents how the obsession with food purity has remained stable in the history of US dietetic advice and continues to generate bifurcation discourses about eating, summed up by George W. Bush's famous post-9/11 dictum that "you're either with us or against us" (DuPuis, 2015, p. 4). As I have documented in the ongoing food wars between "got" and "not" milk claims, pro- and anti-carb enthusiasts, and vegan and carnist divides, there is no shortage of food polarizations propelled by the surprisingly insidious idea that there is only one way to eat and one way to eat well. Like DuPuis (2015) has said, "By imagining a less purified, bounded, and perfectionist relationship to the world, I open up the body—and eating—to the possibility of relational and discursive knowledge, alliances with people in their messy complex selves" (pp. 8–9). While my focus has been on the dangers of singular food truths, DuPuis (2015) upholds the political promise of fermentation as an alternative to food purity discourses.

Decentering anthropocentric understandings of the body as human, she (2015) reminds us that "there is no 'us' without […] microbial others" (p. 138). For her (2015) and others who take up fermentation as an alternative food ontology (see also Hey, 2019), fermentation provides a new metaphor for the complexities of food and eating. It "tells a different story about how to create a more healthy society" (DuPuis, 2015, p. 147)—a story that links us to ecology, microbes and bacteria, human and non-human others, and ultimately to the messy complexities unaccounted for in empirical nutricentric approaches to food and eating. Rather than thinking about food and eating as bound by social discourses of "good" and "bad", "purity" and "danger", "health" and "disease", "the metabiomic body teaches us to think of the world as composed of 'socio-natural-objects', created through the relationships between humans and nature" (DuPuis, 2015, pp. 145–146). When so much of contemporary food politics asks us to unilaterally distinguish between what is true and what is not, "Fermentation enables us to wedge ourselves into a relational food web where we are tethered to others in ecologies of eating" (Hey, 2019, p. 252)—a welcome alternative to the reliance on and maintenance of singular food truths.

King et. al (2019) similarly articulate a framing of "messy eating" to capture the complexity of contemporary eating frameworks. Their commitment to messy eating as a productive food lens is inspired by Alexis Shotwell's suggestion that it is useful to think about "complicity and compromise as a starting point for action" (quoted in King et al., 2019, p. 6). Rejecting any notion of absolute or universal food truths, they contend that messiness as an analytic framework brings "fresh perspectives and new models for living well" outside the individualist underpinnings of dominant conceptualizations of ethical or good eating (p. 9). In place of singular food truths, as I have documented, what we find instead are contradictory, messy, and contextual food truths, which may be seen, on the one hand, as a failure of the empirical model and, on the other, as a more accurate account of the intersectional conditions that define, enable, and constrain "ethical", "pure", "good", and "healthy" eating in the contemporary moment. The analytic lens of messy eating also enables a questioning of who creates and sustains categories of "good", "ethical", "pure", and "healthy" eating. Who benefits? Who is excluded? And to what effect?

Discussed in Chapter 1, food philosopher Lisa Heldke (2012) advances a relational approach to eating that moves beyond a substance-based ontology, questioning how we think about how we should eat. Heldke's (2012) relational food ontology "takes up an ethics that attends to sustaining relationships [...] one that views foods not as substances, but as *loci of relations*" (p. 70, emphasis added). Relation ontologies, including relational eating, are also central for Indigenous scholars. Kim TallBear (2019), a Sisseton Wahpeton Oyate scholar of science studies, draws on the framing of "being in relation" to decentre the animal-human binaries common to Western food discourse. Priscilla Settee (2013), a Cree activist for Native, women's, and environmental rights, articulates the Cree notion of relational well-being (*miýo-pimātisiwin*) that ties not only to the health of food sources, but also to the health of the land, the community, elders, and harmony with nature. Shifting away from the narrow and reductive singular truth frames of Western health discourse, food and well-being in these articulations are marked by a constant state of relatedness and relationality. And I would add that a critical orientation to shifting truth claims is part of thinking, being, and eating *in relation*.

Whether or not we are in the midst of a definitive sea change of food discourse remains to be seen. However, my hope in writing this book has been to advocate for an understanding of food and health beyond the singular truths of nutritionism and to contribute to what seems like growing articulations of relational, contextual, and messy eating already taking place. Until the cultural obsession with singular food truths can be unhinged, we will remain locked in circular, didactic frameworks of food choice that falsely uphold universal truths and negate relational, contextual accounts of food and eating. In broadening the search for contextual, relational truths, I offer an altered way to think about and act upon food, eating, and health beyond the singular, biomedicalized approach that an empirical ontology invariably evokes.

Against singular dietetic advice

In reframing my initial question of what we lose by giving up singular food truths, I also want to explore what we gain from a plural, post-singular-truth food framework. What I hope we gain is an understanding of food that is fluid, shifting, contextual, and contingent. One that shifts in relation to shifting social, political, and ecological contexts. In doing so, we may find we benefit from an appreciation of food that reclaims pleasure, taste, texture, mood, and other embodied, subjective accounts of food and eating—what Mudry, Hayes-Conroy, Chen, and Kimura (2014) refer to as other ways of knowing food. In doing so, we may find we prioritize a model of nutrition that can account for the single largest public health threat, social and economic inequality, by highlighting the structural constraints of food accessibility and affordability. In doing so, we may find we adopt a language of food and eating that reflects its myriad relational aspects. In previous chapters, I have offered examples of how a non-singular food approach can offer insights into the social discourses that regulate contemporary food and eating frameworks, including the individualistic underpinnings of current regulatory maxims of

Alternative Food Facts Table

Serving size: dependent on mood, weather, hunger levels, level of activity, metabolism, pleasure, daily food intake, food pairing, personal digestibility, cultural norms, religious practice.

Affordability	$$
Time needed for preparation	10 – 20 min.
Distance travelled to this vendor	25781 km
Locally sourced	No
Seasonal availability	August - October (USA)
Country of origin	United States
Produced by	Eastern Organics
Treatment of workers	Poor
Work hours	50hrs/week
Hourly wage	$8/hr in USD
Pesticides and herbicides used	None
Genetically modified	No
Vegan	Yes

Figure 6.2 Alternative food facts table[1] (a visual and analytic tool to help work against singular food truths).
Source: Author.

"eating right". As another example of the operationalization of the arguments I am making and how they might be applied as a tool to help work against singular food truths, I offer a mock-up, or reconfiguration, of the traditional nutrition facts table critiqued above (see Figure 6.2).

Using a range and combination of contextualized food values—including, but not limited to, accessibility, activity levels, affordability, availability, connection to community, convenience, cultural and religious practices, digestibility, environmental impact, enjoyment and pleasure, ethical impact, health and nourishment, homemade, locally produced, mood, social norms and discourses, taste, texture, smell, ties to family, weather, and season—the alternative food facts label aims to shift the terms in and through which singular food facts are judged and upheld. A visual and analytic tool, the alternative food facts table is one example of an account of food choice that acknowledges and values relational, contextual, and shifting food truths. The alternative, contextual food facts listed on the label are examples and do not refer to any specific product or company. Moreover, they are not intended to expand the proscriptive approach on how to eat nor are they meant as an exhaustive account of alternative ways to understand food. Indeed, I would encourage others to add or swap out relational, subjective, or contextual factors as they see fit. Depending on the person, day, mood, or other situational factors, a food's accessibility, convenience, and cost may outweigh concerns around health and nourishment. Ties to family and/or community may trump considerations of local production or environmental impact. And cultural and religious practices may offset concerns related to social norms and discourses. Any one of these contextual factors, or any combination thereof, complicates singular food truths because it places the question of what we eat into the complex matrices of day-to-day living and necessarily expands empirical, quantified understandings of food. Amid the many, messy, and increasingly polarized food wars we find ourselves in, breaking away from singular food truths may be one way to reclaim food and eating as deeply subjective and intricately relational, bringing us face-to-face with many of the macro-level issues of our time.

Until dietary advice can be fractured from the false assumption of singular dietary truths, we will continue to blame ill health on poor eating choices, ignore systemic and structural health inequalities, uphold and naturalize Eurocentric food values, and fail to question the cultural frames in and through which discourses of dietary truths continue to circulate and shift. In the many iterations of personal purity and health salvation promised through "good" and "healthy" eating, singular food truths remain steadfast. To borrow what we already know from the longer standing disciplines of gender, race, feminist, colonial, and sexuality studies, empirical frameworks of singular, decontextualized truths have long been dangerous, insufficient, and incomplete. Equally risky, I would add, are singular food truths, especially as they normalize individualized understandings of food and eating. Thus, to reimagine food as relational, contextual, subjective, shifting, and contingent, we must first abandon assumptions of false universals still at work in dominant eating discourse. The enduring question of how to think about food and eating is much more applicable to contemporary questions of health when we abandon the reliance on singular food truths that have become so central to dominant Western framings of health and which have also long led us astray.

Note

1 With thanks to Cressida Heyes for their helpful suggestions in reworking an earlier version of this mock-up.

References

Adichie, C. (2009, October 7). *The danger of a single story* [Video file]. Retrieved from https://www.youtube.com/watch?v=D9Ihs241zeg

DuPuis, E. M. (2015). *Dangerous digestion: The politics of American dietary advice.* Berkeley, CA: University of California Press.

Hayes-Conroy, A., & Hayes-Conroy, J. (2013). *Doing nutrition differently: Critical approaches to diet and dietary intervention.* New York, NY: Routledge.

Heldke, L. (2012). An alternative ontology of food: Beyond metaphysics. *Radical Philosophy Review, 15*(1), 67–88. doi: 10.5840/radphilrev20121518

Hey, M. (2019). Fermentation and the possibility of reimagining relationality. In B. Parker, J. Brady, E. Power, & S. Belyea (Eds.), *Feminist food studies: Intersectional perspectives* (pp. 249–268). Toronto, ON: Women's Press.

King, S., Carey, R. S., Macquarrie, I., Milliousm, V. N., & Power, E. M. (2019). *Messy eating: Conversations on animals as food.* New York, NY: Fordham University Press.

Klein, R. (2010). What is health and how do you get it? In J. M. Metzl & A. Kirkland (Eds.), *Against health: How health became the new morality* (pp. 15–25). New York, NY: New York University Press.

Mayes, C., & Thompson, D. B. (2014). Is nutritional advocacy morally indigestible? A critical analysis of the scientific and ethical implications of "healthy" food choice discourse in liberal societies. *Public Health Ethic, 7*(2), 158–169.

Mudry, J. (2009). *Measured meals: Nutrition in America.* Albany, NY: State University of New York Press.

Mudry, J., Hayes-Conroy, J., Chen, N., & Kimura, A. H. (2014). Other ways of knowing food. *Gastronomica: The Journal of Food and Culture, 14*(3), 27–33. doi: 10.1525/GFC.2014.14.3.27

Settee, P. (2013). *Pimatisiwin: The good life, global Indigenous knowledge systems.* Vernon, BC: J. Charlton Publishing.

Shotwell, A. (2016). *Against purity: Living ethically in compromised times.* Minneapolis, MN: University of Minnesota Press.

TallBear, K. (2019). Being in relation. In S. King, S. R. Carey, I. Macquarrie, V. N. Millious, & E. M. Power (Eds.), *Messy eating: Conversations on animals as food* (pp. 54–67). New York, NY: Fordham University Press.

Index

Note: *Italic* page numbers refer to figures and page numbers followed by "n" denote endnotes.